Cambridge Lower Secondary

Science

STAGE 8: WORKBOOK

Aidan Gill, Heidi Foxford,
Dorothy Warren

Collins

William Collins' dream of knowledge for all began with the publication of his first book in 1819.

A self-educated mill worker, he not only enriched millions of lives, but also founded a flourishing publishing house. Today, staying true to this spirit, Collins books are packed with inspiration, innovation and practical expertise. They place you at the centre of a world of possibility and give you exactly what you need to explore it.

Collins. Freedom to teach.

Published by Collins
An imprint of HarperCollinsPublishers
The News Building
1 London Bridge Street
London
SE1 9GF

Browse the complete Collins catalogue at
www.collins.co.uk

1st Floor, Watermarque Building, Ringsend Road
Dublin 4, Ireland

© HarperCollinsPublishers Limited 2018

10 9 8 7 6 5

ISBN 978-0-00-825472-8

MIX
Paper from
responsible sources
FSC™ C007454

FSC
www.fsc.org

This book is produced from independently certified FSC paper to ensure responsible forest management.

For more information visit:
www.harpercollins.co.uk/green

British Library Cataloguing in Publication Data
A catalogue record for this publication is available from the British Library.

Authors: Aidan Gill, Heidi Foxford, Dorothy Warren
Development editors: Elizabeth Barker, Peter Batty, Richard Needham
Team leaders: Mark Levesley, Peter Robinson, Aidan Gill
Commissioning project manager: Carol Usher
Commissioning editors: Joanna Ramsay, Rachael Harrison
In-house editor: Natasha Paul
Copyeditor: Rebecca Ramsden
Proofreader: Mitch Fitton
Answer checker: Sarah Ryan
Illustrator: Jouve India Private Limited
Cover designer: Gordon MacGilp
Cover illustrations: Maria Herbert-Liew
Internal designer: Jouve India Private Limited
Typesetter: Jouve India Private Limited
Production controller: Tina Paul
Printed and bound in the UK using 100% Renewable Electricity at CPI Group (UK) Ltd

All test-style questions and sample answers in these resources were written by the authors. In Cambridge tests, the way marks are awarded may be different.

Contents

How to use this book

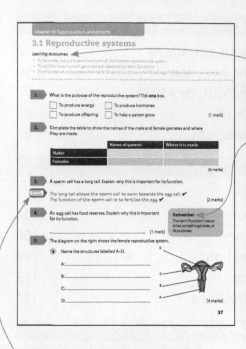

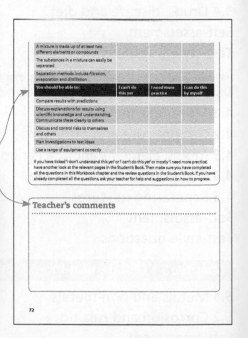

The outcomes show what you will cover in the questions

There are handy hints and tips in the 'remember' boxes

Learn how to structure your answers with 'show me' questions where part of the answer is already completed for you

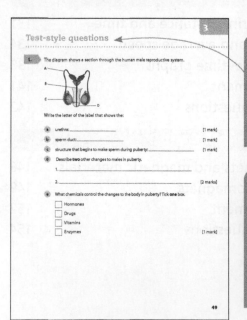

The worked examples will show you a sample answer and how the marks are awarded

Stretch yourself with these challenge questions

There are questions developing your practical skills throughout

At the end of the chapter, try the test-style questions! The questions will cover the topics within the chapter.

At the end of the chapter, fill in the table to work out the areas you understand and the areas where you might need more practice. There is space for your teacher to comment too

Biology

1.1 Food from plants

Learning outcomes

- To describe how plants make their own food, using photosynthesis
- To explain why plants need a continuous source of water
- To explain how leaves are adapted for photosynthesis

1. In which organisms does photosynthesis happen? Tick **one** box.

☐ Plants

☐ Animals

☐ Plants and animals

☐ None of the above [1 mark]

2. List **two** raw materials (reactants) that plants need for photosynthesis.

1. _____

2. _____ [2 marks]

3. Explain why photosynthesis does **not** happen at night.

Worked Example

Plants cannot photosynthesise at night because there is no light at
night ✔ and plants need light to be able to photosynthesise. ✔ [2 marks]

4. Khadija is investigating photosynthesis in pondweed.

Practical

- lamp
- test tube
- gas
- water
- funnel
- pondweed

> **Remember**
> The command word 'explain' means you need to give the reason why something does or does not happen.

Bubbles of gas are produced.

a What gas is produced? _____ [1 mark]

b How could Khadija test for the gas? _____ [1 mark]

5. From what is most of a plant's biomass made? Tick **one** box.

☐ Soil

☐ Carbon dioxide and water

☐ Nutrients (from the soil)

☐ Water only

[1 mark]

Remember

Biomass is the total dry mass of an organism. As plants photosynthesise and grow, their biomass increases.

6. The diagram below shows a section through a leaf.

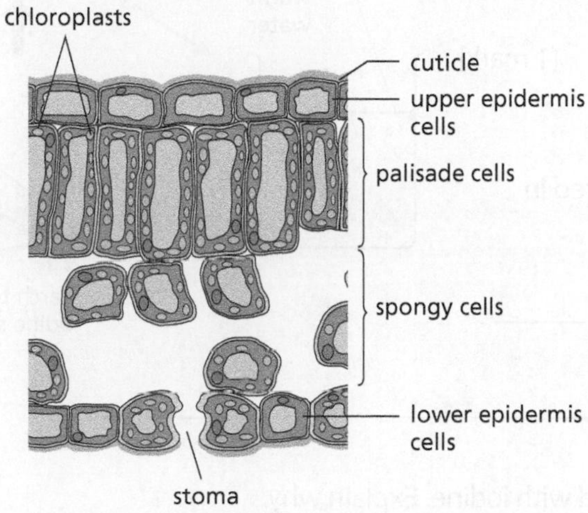

Use the diagram to help you answer the questions.

a In which part of the leaf does most photosynthesis take place?

_____ [1 mark]

b Which part of the leaf has air spaces so that gases can move around?

_____ [1 mark]

c Through which part does water escape?

_____ [1 mark]

7. Some trees have small needles instead of leaves. The needles have fewer stomata and a thick waxy cuticle.

Show Me

Explain **one** way this helps the tree survive.

Fewer stomata and a thick waxy cuticle help the tree lose less _____ .

This is important for survival because water is needed for _____ . [2 marks]

8.

Practical

Rashid is testing a plant leaf for starch.
The diagram shows the method used.

leaf
boiling water
heat
ethanol
hot water
heat
warm water
leaf is washed

starch test with iodine solution

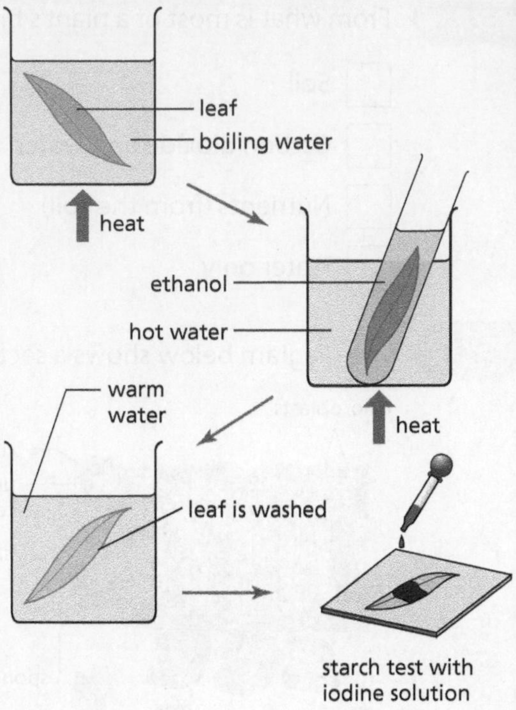

a What idea is Rashid testing? Tick **one** box.

☐ That plants produce starch when they photosynthesise

☐ That plants absorb water

☐ That plants contain chlorophyll

☐ That plants contain cells   [1 mark]

b Explain why the leaf needs to be heated in ethanol.

_____

_____ [1 mark]

c The leaf turns blue-black when tested with iodine. Explain why.

_____

_____

_____ [2 marks]

d Rashid tests a leaf that has been covered in foil for 48 hours.

Predict what will happen when he tests this leaf for starch.

_____

_____ [1 mark]

e Explain your prediction.

_____

_____

_____ [2 marks]

9.

Surika has a variegated leaf from a plant that has been in a sunny place for two days. A variegated leaf has both green and white parts.

She boils the leaf in water.

Then she then puts it in hot ethanol.

Finally, she tests it with iodine solution.

Complete the table to show what results you would expect and why.

	Green part of variegated leaf	White part of variegated leaf
Result of starch test using iodine		
Explanation		

[4 marks]

1.2 A balanced diet

Learning outcomes

- To describe what a balanced diet contains
- To explain why our bodies need carbohydrates, fats, proteins, vitamins, minerals, fibre and water
- To describe how diet and exercise affect our bodies

1.

Draw a line to match each food group with the reason it is needed.

Food group	Reason food group is needed
Protein	To store energy
Carbohydrate	For growth and repair
Fat	For energy

[2 marks]

2. Name the main food group we get from:

a fish _____ [1 mark]

b rice _____ [1 mark]

3. List **two** ways the body uses energy from food.

1. _____ 2. _____ [2 marks]

4. Which of the following foods is a good source of fibre? Tick **one** box.

☐ Yoghurt ☐ Eggs ☐ Lentils ☐ Chicken [1 mark]

5. Obesity is a condition linked to what sort of diet? Tick **one** box.

☐ A diet high in fat and high-energy foods

☐ A diet low in fat

☐ A diet high in fibre

☐ A balanced diet [1 mark]

6. Osama is burning different types of nut to find out how much energy is released.
He uses the apparatus shown below.

Practical

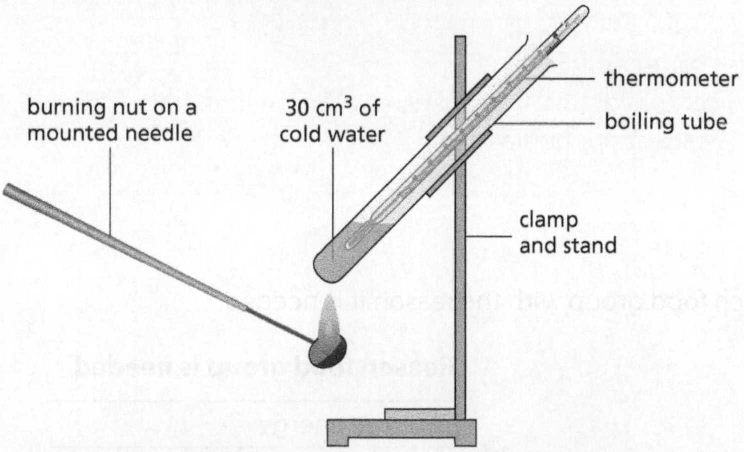

burning nut on a mounted needle   30 cm³ of cold water   thermometer   boiling tube   clamp and stand

Osama burns 2 grams of each type of nut. He measures the increase in the temperature of the water in the boiling tube.

Show Me

a Write down **two** safety precautions Osama needs to take to reduce the risks to himself.

Keeping clothes away from _____ .

Do not touch _____ . [2 marks]

10

Osama's results are shown in the table.

Type of nut	Temperature of water at start (°C)	Temperature of water at end (°C)	Rise in temperature (°C)
Pecan	16.3	26.8	10.5
Almond	17.2	21.5	
Cashew	17.7	23.7	

b Complete the table by calculating the change in the temperature for the almond and cashew nut. [2 marks]

c In Osama's investigation state:

The variable being changed: _____ [1 mark]

The variable being measured: _____ [1 mark]

d Which type of nut contained the most energy? _____ [1 mark]

7. Fatma is an Olympic athlete. Her friends suggest she should only eat foods very high in protein such as chicken, soya and eggs.

a Give **one** reason why an athlete needs extra protein in their diet.

_____

_____ [1 mark]

b Explain why Fatma should **not** only eat foods very high in protein.

_____

_____

_____

_____ [2 marks]

8. The table below shows the nutrition facts for a breakfast cereal for children.

	Per 100 g	Per 30 g serving
Energy	1647 kJ (389 kcal)	494 kJ (117 kcal)
Fat	2.8 g	0.8 g
Carbohydrates	84 g	25 g
of which sugars	32 g	9.6 g
Fibre	2.3 g	0.7 g
Protein	6.2 g	1.9 g
Salt	0.75 g	0.22 g

a Explain why food labels show amounts per 100 g.

_____ [1 mark]

b The food labels say that the cereal contains protein. Give the name of a food test that could test for protein. _____ [1 mark]

Challenge **c** Use evidence from the table to evaluate whether this cereal is healthy.

_____

_____

_____ [2 marks]

9. The table shows how many kilojoules of energy are required for a 60 kg person to do different activities for **one** hour.

Activity	kJ
Standing in line	300
Playing cricket	1200
Playing football	1740
Running up stairs	3780

Use the information in the table to calculate:

a the number of kJ (kilojoules) needed for a 60 kg person to play cricket for **two** hours.

_____ [1 mark]

b   the number of kJ (kilojoules) needed for a 60 kg person to play football for **two** hours.

_____ [1 mark]

c   Explain why a football player might need to eat more carbohydrates than a cricket player.

_____

_____

_____ [2 marks]

1.3 Malnutrition

Learning outcomes
- To describe what happens if you lack nutrients in your diet
- To explain why a good diet is important for fitness

1.   What is malnutrition? Tick **one** box.

☐   When a diet contains too much or too little of something, and causes problems

☐   When a diet is too low in calories

☐   When somebody overeats and becomes obese

☐   When somebody is too thin   [1 mark]

2.   Which statement best describes fitness? Tick **one** box.

☐   Living to an old age

☐   Being able to run very fast

☐   Being able to do everyday tasks without being out of breath

☐   Doing exercise all the time   [1 mark]

3.   Draw **one** line to match each nutritional deficiency with the effect it can have on the body.

Nutritional deficiency	Potential effect on body
Lack of iron	Rickets
Lack of vitamin D	Kwashiorkor
Lack of protein	Anaemia

[2 marks]

4. Which **two** of the following conditions are often caused by overeating fatty foods?

☐ Low blood pressure   ☐ Obesity

☐ High blood pressure   ☐ Scurvy   [2 marks]

5. Name a condition that results from too little fibre in the diet. _____ [1 mark]

6. The table shows the recommended amounts of iron for different groups of people.

Group	Age (years)	Iron (mg/day)
Males	19–50	8
Females	19–50	18
Pregnant females	19–50	27

a Which group of people need least iron? _____ [1 mark]

b A 30-year-old pregnant female has a diet containing 19 mg/iron a day. Is this above

or below the recommended dietary allowance? _____ [1 mark]

c Suggest a reason why pregnant women need more iron than non-pregnant women.

_____

_____ [1 mark]

7. Vegetarians can get all the nutrients they need if they have a balanced diet.
Explain why this is.

Show Me

A balanced diet contains _____ .

A balanced diet can therefore provide all the _____
a vegetarian needs.   [2 marks]

8. Water is not a nutrient but we need water to survive. Explain why.

Challenge

_____

_____

_____

_____ [2 marks]

Remember
Approximately 70% of the body is water. The chemical reactions in our cells take place in water.

9. In 1887, a nerve disease attacked people in the Dutch East Indies. The disease was beriberi. Symptoms included weakness and loss of appetite, and victims often died of heart failure.

Scientists thought the disease might be caused by bacteria. They injected chickens with bacteria from the blood of patients with beriberi. The injected chickens became sick. However, so did a group of chickens that were not injected with bacteria.

Dr Eijkman noticed that before the experiment all the chickens ate whole-grain rice; but during the experiment they ate polished rice. He researched this and found that polished rice lacked thiamine, a vitamin necessary for good health.

a What idea were the scientists testing when they injected the chickens with blood of patients with beriberi?

_____ [1 mark]

b What conclusion would you make from the experiment?

_____

_____ [1 mark]

1.4 Digestion

Learning outcomes

- To identify the organs of the alimentary canal
- To describe the functions of the organs involved in digestion
- To describe what happens in digestion, including the action of enzymes

1. The diagram shows the human digestive system.

1: Name organs A, B and C.

A: _____

B: _____

C: _____ [1 mark]

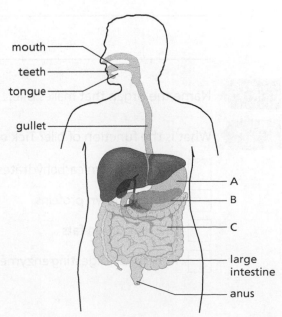

2. Where does digestion begin?

_____ [1 mark]

3. Describe **two** functions of saliva.

To make food easier _____ and to

digest _____ . [2 marks]

4. What word is used to describe how food is moved down the oesophagus (gullet)?

_____ [1 mark]

> **Remember**
> The command word 'describe' means to write about what something is like.

5. What happens in the stomach? Tick **one** box.

☐ Mechanical digestion

☐ Chemical digestion

☐ Mechanical and chemical digestion

☐ Neither mechanical nor chemical digestion [1 mark]

6. Complete the sentences using words from the list.

nervous	large intestine	circulatory	blood	digestive	pancreas

Nutrients pass from the small intestine into the _____ .

They are then transported around the body by the _____ system. [2 marks]

7. Digestive juices contain enzymes. Explain the function of digestive enzymes.

_____

_____ [2 marks]

8. Name the organ that makes bile. _____ [1 mark]

9. What is the function of bile? Tick **one** box.

☐ To break down carbohydrates

☐ To break down proteins

☐ To break down fats

☐ To help fat-digesting enzymes to work [1 mark]

10. Describe what happens in the large intestine.

_____

[1 mark]

11. Compare mechanical and chemical digestion.

_____

_____

[2 marks]

12. Enzymes are biological catalysts. Explain what is meant by 'biological catalyst'.

Challenge

_____

_____

_____

[2 marks]

Self-assessment

Tick the column which best describes what you know and what you are able to do.

What you should know:	I don't understand this yet	I need more practice	I understand this
Photosynthesis is a chemical reaction in which carbon dioxide and water turn into glucose and oxygen			
Chloroplasts contain green chlorophyll, which traps light for photosynthesis			
A plant uses glucose and mineral salts (from the soil) to make all its compounds			
The mass of the compounds that a plant has made for itself is its biomass			
Plants make starch to store energy			
We test for starch using iodine solution, which turns from orange to blue-black			
Your diet needs to contain nutrients (carbohydrates, fats, proteins, minerals and vitamins) for energy, growth and repair, and health			
You need water and fibre in your diet too			

	I can't do this yet	I need more practice	I can do this by myself
Carbohydrates and fats store energy, which is measured in joules (J) or kilojoules (kJ)			
Eating too much carbohydrate or fat may cause weight gain and make people unhealthy			
We use tests to discover which foods contain which nutrients			
For a balanced diet you need to eat foods from many sources			
Malnutrition is when a diet contains too much or too little of something, which then causes health problems			
Malnutrition causes deficiency diseases (such as kwashiorkor and scurvy)			
Malnutrition also causes obesity, type 2 diabetes and high blood pressure			
You need to digest food for it to be useful inside your body			
The digestive system contains the alimentary canal, and some other organs (such as the liver, pancreas and salivary glands)			
The alimentary canal is a series of organs that form a tube from your mouth to your anus			
Digestive juices contain enzymes, which act as biological catalysts to digest food			

You should be able to:	I can't do this yet	I need more practice	I can do this by myself
Select ideas and turn them into a form that can be tested			
Make predictions using scientific knowledge and understanding			
Discuss and control risks to yourself and others			
Use a range of equipment correctly			
Identify important variables			
Make simple calculations			
Interpret data from secondary sources			
Discuss explanations for results using scientific knowledge and understanding			
Present results as appropriate in tables			

If you have ticked 'I don't understand this yet' or 'I can't do this yet' or mostly 'I need more practice', have another look at the relevant pages in the Student's Book. Then make sure you have completed all the questions in this Workbook chapter and the review questions in the Student's Book. If you have already completed all the questions, ask your teacher for help and suggestions on how to progress.

Teacher's comments

....

Test-style questions
....

1. Dinesh is investigating what nutrients are found in bread.

He adds a drop of iodine solution onto a piece of bread.

a Complete the prediction using the words from the list:

stay orange-brown	turn blue-black	turn pink

The iodine will _____ because the bread contains starch. [1 mark]

b Iodine can be harmful. Write down **one** way in which Dinesh can control the risks to himself.

_____ [1 mark]

c Describe **one** way the body uses the starch we eat.

_____ [1 mark]

d Name **one** condition that can result from eating too many starchy foods.

_____ [1 mark]

2. Water lily plants live in water.

a Name **two** substances the water lily always needs to make food.

1. _____

2. _____ [2 marks]

b Name the chemical reaction that plants use to make food.

_____ [1 mark]

Mia tests a lily leaf to find out if it contains starch. These are the steps she uses. They are in the **incorrect** order.

A	B	C	D
Wash the leaf in water (to remove ethanol and soften it again)	Add iodine solution to leaf	Remove chlorophyll using ethanol	Soften the leaf by boiling in water

c Write the letters to show the correct order.

_____ [1 mark]

d Water lilies have large flat leaves covering the surface of the water.

Explain how this feature helps the plant survive.

_____

_____

_____ [2 marks]

e Suggest why you find few plants growing underneath lilies.

_____

_____

_____ [2 marks]

f Explain how a water lily is different to a human in the way it gets its nutrition.

_____

_____

_____ [2 marks]

3. Malik is researching how carbohydrates are digested.

a State why we need carbohydrates in our diet.

_____ [1 mark]

b The diagram shows part of the alimentary canal.

On the diagram, label the following parts:

- the stomach
- the small intestine
- the large intestine.

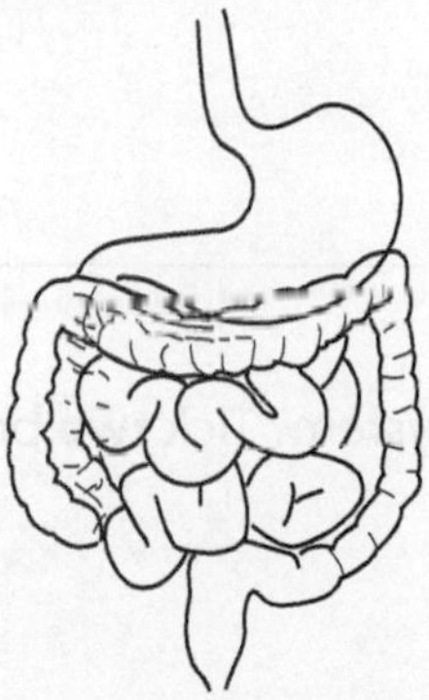

[3 marks]

c Explain what happens to carbohydrates in the small intestine.

_____

_____

_____ [2 marks]

d Explain why an athlete needs to eat more carbohydrates.

_____

_____

_____ [2 marks]

2.1 Human circulatory system

Learning outcomes

- To describe the parts and functions of the circulatory system
- To explain how oxygen and food substances are delivered to all cells
- To describe some circulatory system disorders

1. Complete the sentences using words from the list.

blood	cells	circulation	respiration	vessels	water

The circulatory system carries _____ around your body in tubes called

blood _____ . The movement of blood around your body is called

your _____ . [3 marks]

2. Why do cells in the body need oxygen?

_____ [1 mark]

3. Which of the following organs are part of the circulatory system? Tick **two** boxes.

☐ The stomach

☐ The lungs

☐ Blood vessels

☐ The heart [2 marks]

4. Describe the main function of the heart.

_____ [1 mark]

5. Complete the table to show the names and functions of the types of blood vessels.

Name of vessel	Function
Arteries	
Veins	
	Supply cells with substances they need from the blood.

[3 marks]

6. Describe how eating too many fatty foods can affect the arteries.

Worked Example

Fatty substances form plaques inside the arteries. ✔
These make the arteries narrower. ✔   [2 marks]

7. Explain what happens if muscle tissue in the heart does **not** get enough blood.

Show Me

Muscle tissue will not get enough

_____ so cells will not be able to

respire and _____   [2 marks]

Remember
Questions that are worth two marks will need two different points to gain the two marks.

8. The diagram shows part of the circulatory system.

Write down the letter of the vessel that:

a carries blood that is higher in oxygen.

_____   [1 mark]

b has thick and strong walls

_____   [1 mark]

c carries blood that is lower in oxygen

_____   [1 mark]

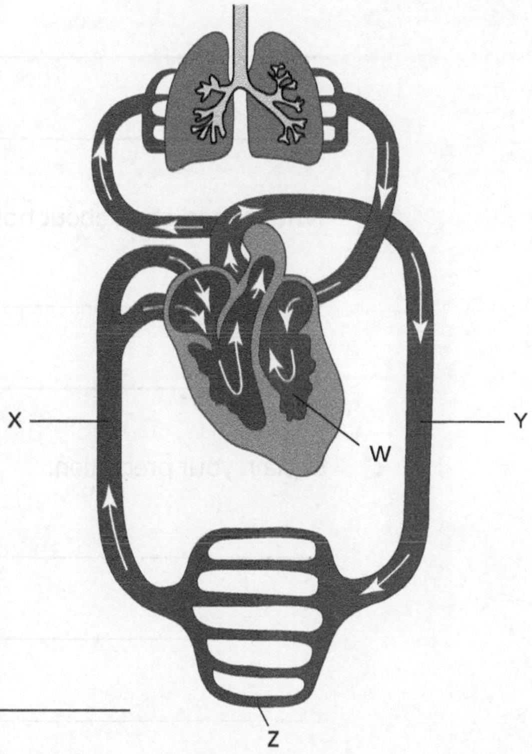

X ———   ——— Y

W

Z

9. In the diagram for question 8, the wall of chamber W is thicker than the chamber on the other side of the heart. Explain why this is.

_____

_____

_____   [2 marks]

10. Explain why the circulatory system is described as a double circulatory system.

Challenge

_____

_____

_____   [2 marks]

11.

Worked Example

Amina takes her pulse. She counts 19 beats in 15 seconds.

Calculate her pulse rate in beats per minute (bpm).

First, calculate how many lots of 15 seconds are in 60 seconds: $60 \div 15 = 4$.

This means you must multiply the number of beats by 4 to get the number of beats per minute.

Therefore, $19 \times 4 = 76$ bpm. ✔   [1 mark]

12.

Practical

a Describe how Amina could test how her pulse is affected by exercise.

_____

_____

_____   [2 marks]

b Write a prediction about how you think Amina's pulse rate will be affected by exercise.

_____

_____   [1 mark]

c Explain your prediction.

_____

_____

_____   [2 marks]

13. Draw **one** line to match the part of the blood with its function.

Part of blood	Function
Plasma	Helps blood to form clots
Platelets	Transports carbon dioxide, digested food and urea in the blood
Red blood cells	Transports oxygen in the blood

[2 marks]

2.2 Human respiratory system

Learning outcomes

- To model aerobic respiration using a word equation
- To explain how oxygen enters the blood, and how carbon dioxide is removed
- To describe the parts and functions of the respiratory system

1. The diagram shows the parts of the human respiratory system.

Name the parts labelled A, B and C.

A: _____

B: _____

C: _____   [3 marks]

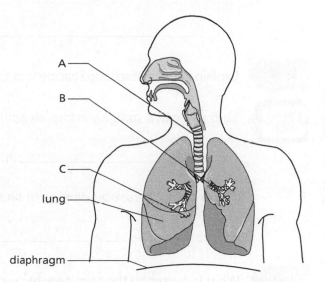

2. Complete the sentences using words from the list.

| outwards | inwards | contract | relax | decreases | increases |

When we inhale, muscles between the ribs _____ and move the ribcage

upwards and _____ . Muscles in the diaphragm contract and flatten it.

This _____ the volume of the chest.   [3 marks]

3. Oliver measures his breathing rate while running. Here are his results.

Time (minutes)	Breathing rate (breaths per minute)
0	12
1	22
2	24
3	26
4	28
5	30

a What variable does he change? _____ [1 mark]

b What variable does he measure? _____ [1 mark]

c Suggest an explanation for Oliver's results.

_____

_____

_____ [2 marks]

4. Explain how gas exchange happens in the alveoli.

Show Me

There is more oxygen in the alveoli than in the capillaries so oxygen moves

from the _____ into the _____ .

There is more carbon dioxide in the capillaries than in the lungs so carbon dioxide

moves from the _____ into the _____ . [2 marks]

5. What is meant by the term **aerobic** respiration?

_____ [1 mark]

6. Where does aerobic respiration take place? Tick **one** box.

☐ In muscles   ☐ In the heart

☐ In the lungs   ☐ In all cells   [1 mark]

7. Complete the word equation for aerobic respiration.

_____ + oxygen ⟶ carbon dioxide + _____

[2 marks]

Remember

Energy is not written as a product of respiration as it is not a substance.

8. Explain the difference between respiration and breathing.

_____

_____

_____ [2 marks]

9.

Challenge

Long distance cyclists often drink liquids that contain glucose.
Explain how this helps them perform well.

_____

_____

_____ [2 marks]

10.

Practical

Yuri makes a model of the lungs as shown below.

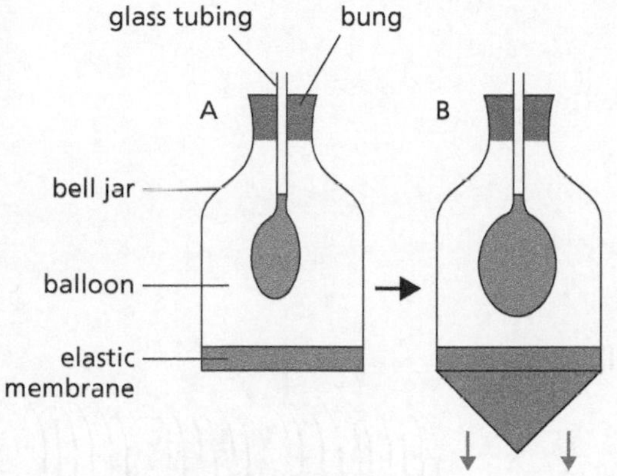

glass tubing   bung

A   B

bell jar

balloon

elastic
membrane

State what is represented by:

a   The elastic membrane _____   [1 mark]

b   The bell jar _____   [1 mark]

c   The balloon _____   [1 mark]

d   When Yuri pulls down on the rubber sheet, the balloon gets bigger. Explain why.

_____

_____

_____ [3 marks]

e   Predict what would happen if he pushed the rubber sheet upwards. Explain your prediction.

_____

_____

_____ [3 marks]

2.3 Smoking and health

Learning outcomes
- To describe how your body keeps your lungs clean
- To explain why your lungs need to be kept clean
- To describe some of the problems caused by smoking

1. How does the nose help keep the lungs clean? Tick **one** box.

☐ It contains hairs and mucus

☐ It contains white blood cells

☐ It secretes antibacterial substances

☐ It secretes enzymes [1 mark]

2. The diagram shows a type of specialised cell found in the trachea.

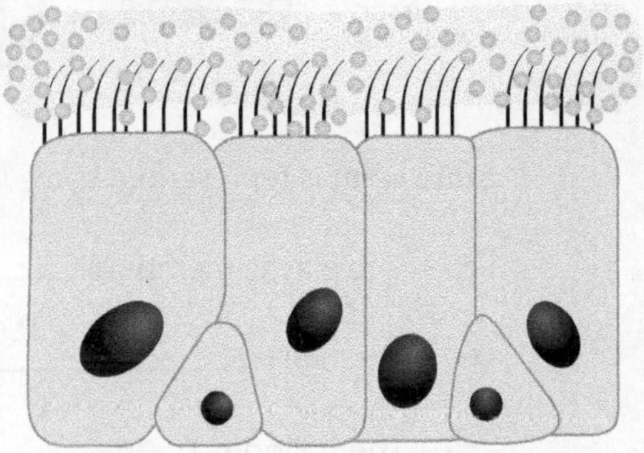

a Name the tall cells shown in the diagram.

_____

[1 mark]

b Describe the function of these cells.

_____ [1 mark]

c Describe the effect of cigarette smoke on these cells.

_____ [1 mark]

3. Name a substance found in cigarette smoke that:

a is addictive: _____ [1 mark]

b can cause cancer: _____ [1 mark]

4. The graph below shows the correlation between the number of cigarettes smoked per day and the number of deaths from lung cancer (per 100 000 men per year).

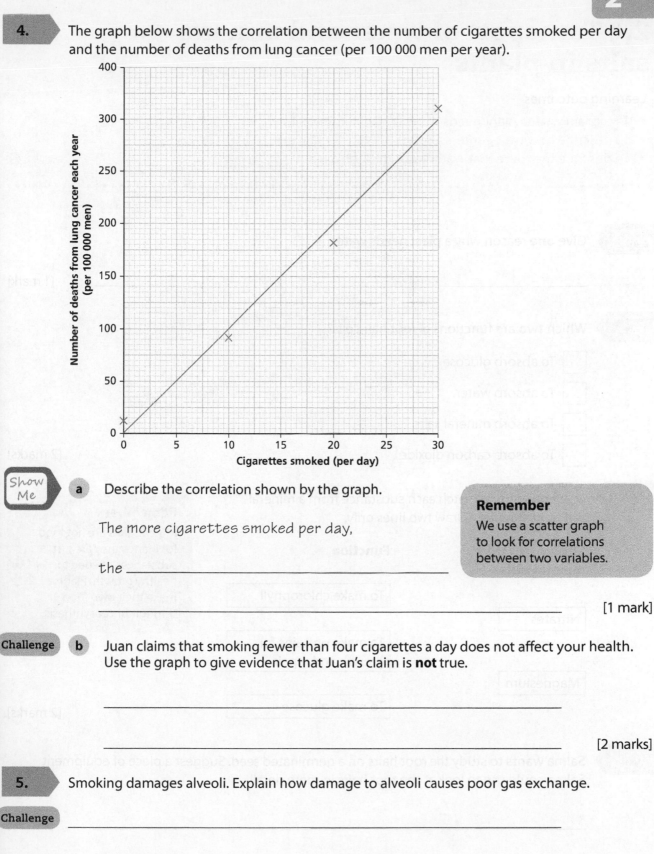

Show Me **a** Describe the correlation shown by the graph.

The more cigarettes smoked per day,

the _____

_____

[1 mark]

> **Remember**
> We use a scatter graph to look for correlations between two variables.

Challenge **b** Juan claims that smoking fewer than four cigarettes a day does not affect your health. Use the graph to give evidence that Juan's claim is **not** true.

_____

_____

[2 marks]

5. Smoking damages alveoli. Explain how damage to alveoli causes poor gas exchange.

Challenge

_____

_____

_____

[2 marks]

2.4 Transport of water and mineral salts in plants

Learning outcomes

- To explain what a plant needs water and mineral salts for
- To describe how water enters a plant
- To describe the route that water takes through a plant

1. Give **one** reason why a plant needs water.

_____ [1 mark]

2. Which **two** are functions of root hair cells?

☐ To absorb glucose

☐ To absorb water

☐ To absorb mineral salts

☐ To absorb carbon dioxide

[2 marks]

3. Draw **one** line to match each substance from a mineral salt to its function. Draw two lines only.

Substances from mineral salts	Function
	To make chlorophyll
Nitrates	To make proteins for growth and repair
Magnesium	To make glucose

[2 marks]

> **Remember**
> Mineral salts are not food for plants, they are just substances needed for healthy growth. Plants make their own food through photosynthesis.

4. Salma wants to study the root hairs on a germinated seed. Suggest a piece of equipment Salma could use to look at the root hairs in more detail.

_____ [1 mark]

5. The diagram shows a root hair cell.

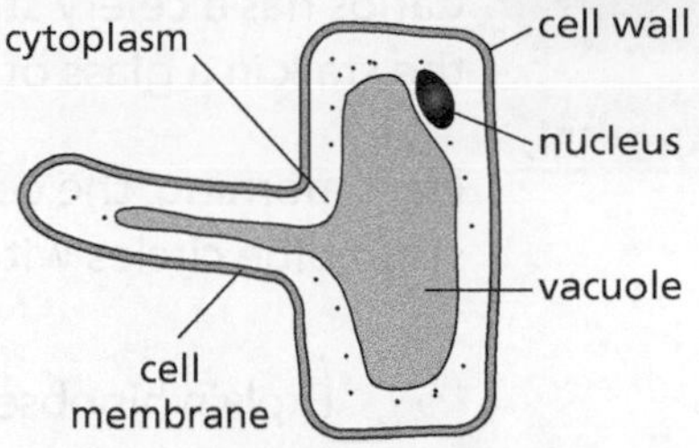

Explain **one** way that the root hair cell is adapted to its function.

Show Me

Its shape gives it a large _____ so it

can absorb _____ more efficiently.

cytoplasm

cell wall

nucleus

vacuole

cell membrane

[2 marks]

6. Gabriella forgets to water her houseplant for three weeks. The leaves become floppy. Explain why.

_____

_____ [2 marks]

Remember

Questions using the command word 'explain' will usually need an answer that contains the word 'so' or 'therefore' or 'because'.

7. The stem of a plant contains xylem tubes. Name **two** substances transported in the xylem.

1._____

2._____ [2 marks]

8. Pierre puts the stem of a white flower in a pot of red food colouring mixed with water. After 24 hours the white flower turns pink. Explain why.

_____

_____ [2 marks]

9. Xylem cells have thick walls stiffened with lignin. Explain how this feature makes xylem adapted to transporting substances.

_____

_____ [2 marks]

10. Why does water move up a stem more quickly on a hot day than on a cool day?

Challenge

_____

_____

_____ [2 marks]

11.

Practical

Carlos has a celery stalk with leaves. He cuts the bottom off the stalk. Then he puts the stalk in a glass of water mixed with blue food colouring.

Next morning, the celery leaves are blue. Carlos cuts the stalk halfway up and sees small blue circles within the stalk.

a   Explain his observations.

_____

_____

_____   [2 marks]

b   What are the blue circles inside the stalk?

_____

_____   [1 mark]

c   Why have the leaves turned blue?

_____

_____   [1 mark]

d   Carlos's friend says that if you cut the leaves off a celery stalk, less water will be taken in by the stalk.

Plan an investigation method to find out if this might be true.

_____

_____

_____

_____   [4 marks]

Self-assessment

Tick the column which best describes what you know and what you are able to do.

What you should know:	I don't understand this yet	I need more practice	I understand this
The human circulatory system contains the heart, blood and blood vessels (arteries, veins, capillaries)			
The circulatory system ensures that all cells have enough oxygen and food, and removes waste products			
Blood plasma carries dissolved food substances, and waste products (such as urea and carbon dioxide)			
Red blood cells carry oxygen			
White blood cells destroy microorganisms			
Platelets help the blood to form clots			
Disorders of the circulatory system occur when fatty plaques make arteries so narrow that tissues do not get enough blood			
The respiratory system contains many organs, including the lungs, diaphragm and trachea			
Breathing is the movement of muscles in the respiratory system			
Breathing changes the pressure inside the chest, which causes air to enter or leave the lungs			
Gaseous exchange occurs when oxygen diffuses from the alveoli into the blood and carbon dioxide diffuses in the opposite direction			
Aerobic respiration can be shown using a word equation: oxygen + glucose $\rightarrow$ carbon dioxide + water			
Specialised cells keep your lungs clean (such as cells that produce mucus and ciliated epithelial cells)			
Tobacco smoke paralyses cilia, and contains an addictive drug called nicotine			
Tar in tobacco smoke causes cancer			

	I can't do this yet	I need more practice	I can do this by myself
Plants need water to make their own food by photosynthesis and for cells to keep their shapes			
A plant needs mineral salts to make substances (such as proteins)			
Water and mineral salts are absorbed by a plant using root hair cells			
Water and mineral salts are transported in tubes formed by dead xylem cells			
Water is lost from a plant through its stomata			
You should be able to:	**I can't do this yet**	**I need more practice**	**I can do this by myself**
Plan investigations to test ideas			
Make predictions using scientific knowledge and understanding			
Be able to suggest explanations for results using scientific knowledge and understanding			
Be able to identify variables in an investigation			
Interpret data from secondary sources			
Identify trends and patterns in results (correlations)			
Identify anomalous results			

If you have ticked 'I don't understand this yet' or 'I can't do this yet' or mostly 'I need more practice', have another look at the relevant pages in the Student's Book. Then make sure you have completed all the questions in this Workbook chapter and the review questions in the Student's Book. If you have already completed all the questions, ask your teacher for help and suggestions on how to progress.

Teacher's comments

Test-style questions

1. The diagram on the right shows the ribcage.

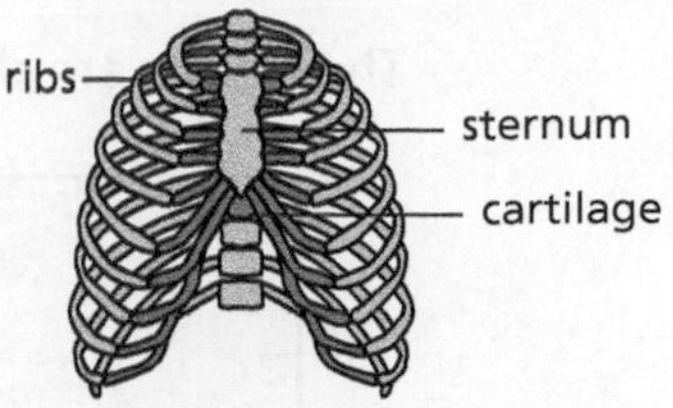

ribs
sternum
cartilage

a Name an organ protected by the ribcage that pumps blood.

_____ [1 mark]

b Name an organ found in the ribcage that is the site of gaseous exchange.

_____ [1 mark]

c Cartilage joins each rib to the sternum. This allows the ribs to move.

Explain why it is important that the ribs can move.

_____

_____ [2 marks]

d The airways found inside the ribcage are kept clean by ciliated epithelial cells.

Explain what happens to these cells in a person that smokes.

_____

_____

_____ [2 marks]

2. The diagram on the right shows two blood vessels.

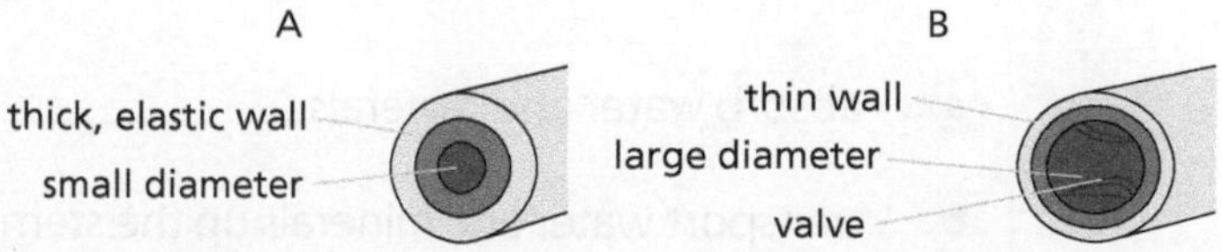

A
thick, elastic wall
small diameter

B
thin wall
large diameter
valve

a Name each of the vessels shown in the diagram.

A: _____   B: _____ [2 marks]

b Explain why vessel **A** has thick elastic walls.

_____

_____ [2 marks]

3. Some scientists collected data to investigate the link between height and lung volume.

They measured the height and lung volume of a group of men aged between 20 and 30.

The results are shown below.

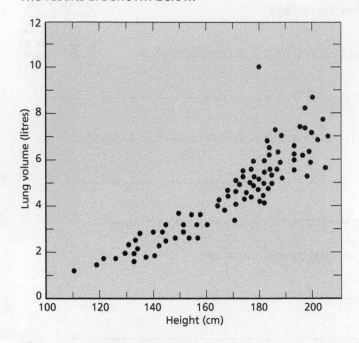

a Describe what this data shows about the correlation between height and lung volume.

_____ [1 mark]

b One man in the investigation was an Olympic athlete. How tall was this man?

_____ [1 mark]

4. Hussain is growing plants. He waters them every day. Once a week he adds fertiliser to the soil.

Name the specialised cells in the plants that:

a absorb water and minerals _____ [1 mark]

b transport water and minerals up the stem to the leaves. _____ [1 mark]

c If plants do not get enough nitrates they don't grow properly. Explain why.

_____

_____ [2 marks]

d Hussain removes the roots from a plant. Explain why the plant could die after the roots are removed.

_____

_____ [2 marks]

3.1 Reproductive systems

Learning outcomes

- To describe the parts and functions of the human reproductive system
- To explain how human gametes are adapted for their functions
- To describe what happens during fertilisation and how a fertilised egg cell develops into an embryo

1. What is the purpose of the reproductive system? Tick **one** box.

   ☐ To produce energy   ☐ To produce hormones

   ☐ To produce offspring   ☐ To help a person grow   [1 mark]

2. Complete the table to show the names of the male and female gametes and where they are made.

	Name of gamete	Where it is made
Males		
Females		

[4 marks]

3. A sperm cell has a long tail. Explain why this is important for its function.

Worked Example

The long tail allows the sperm cell to swim towards the egg cell. ✔
The function of the sperm cell is to fertilise the egg. ✔   [2 marks]

4. An egg cell has food reserves. Explain why this is important for its function.

_____ [1 mark]

Remember

The term 'function' means what somethings does, or its purpose.

5. The diagram on the right shows the female reproductive system.

   a   Name the structures labelled A–D.

   A: _____

   B: _____

   C: _____

   D: _____

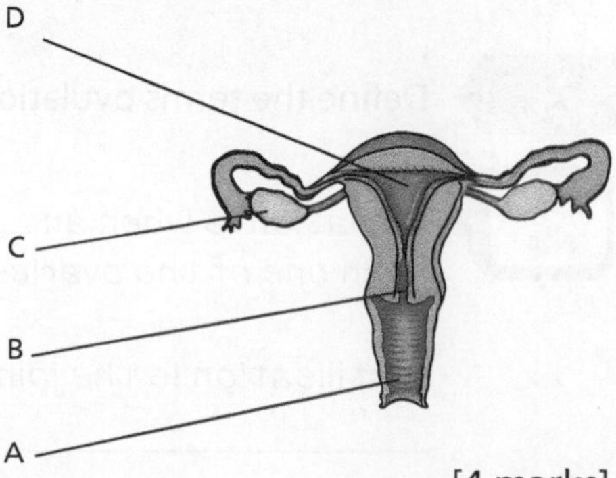

[4 marks]

b Add a label to the diagram to show an oviduct. [1 mark]

c Which letter shows the structure where a developing foetus grows?

_____ [1 mark]

6. Draw a line to match each part of the male reproductive system to its function.

Part of male reproductive system	Function
Sperm duct	Where sperm is made
Urethra	The tube that carries sperm from a testis to the urethra during ejaculation
Testis	The tube that carries urine or sperm to the tip of the penis

[2 marks]

7. Complete the sentence using the words from the list.

ovary	penis	vagina	oviduct	uterus	scrotum	bladder

Sperm is ejaculated from the _____ of the male into the top of the

_____ of the female. From here the sperm go through the cervix, and swim

into the uterus to the _____ . If an egg cell has recently been released from

the _____ the two cells may join. [4 marks]

8. Why is fertilisation in humans called 'internal fertilisation'?

_____

_____ [1 mark]

9. Define the terms ovulation and fertilisation.

Show Me

Ovulation is when an _____ is released
from one of the ovaries into the oviduct.

Fertilisation is the joining of the nuclei from the

_____ to form a fertilised egg.

> **Remember**
> It is important to learn the meaning of all the scientific words in each topic. You should use scientific words in your answers.

[2 marks]

10.

Suggest why many more sperm cells are produced than egg cells.

_____

_____ [1 mark]

11. Some scientists investigated the number of eggs in the ovaries of women of different ages. The graph shows their results.

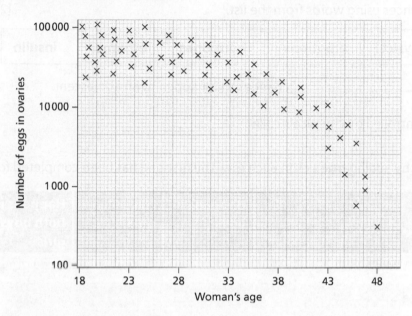

a On average, in which age range do women have most eggs in their ovaries? Tick **one** box.

☐ 18–28

☐ 28–38

☐ 38–48

☐ They have the same amount of eggs in all the age ranges [1 mark]

b Describe the pattern shown by the graph.

_____

_____

_____

_____

> **Remember**
> If you are asked to describe a pattern or trend in data you need to use words such as older/younger, longer/shorter, warmer/cooler, depending on the data shown.

[1 mark]

3.2 Puberty

Learning outcomes

- To describe the stages of human growth and development
- To describe what happens during puberty and adolescence
- To describe what happens in the menstrual cycle

1. Complete the sentences using words from the list. [2 marks]

chemical	physical	emotional	hormones	enzymes	insulin

Puberty is the _____ changes that happen to an adolescent.

The changes are controlled by chemicals called _____ .

2. Complete the table by adding **one** tick in each row. The first one has been completed for you.

Event in puberty	Happens in boys	Happens in girls	Happens in both boys and girls
Faster growth			✔
Deepening of voice			
Increase in underarm body hair			
Widening of hips			

[3 marks]

3. The graph shows the menstrual cycle.

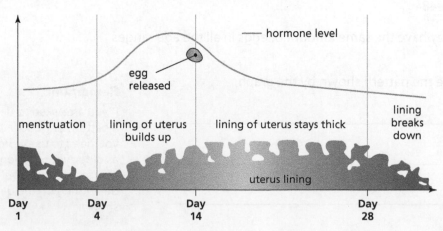

Use the graph to answer the questions:

a On what day does the lining of the uterus begin to build up?

_____ [1 mark]

b On what day does ovulation occur? _____ [1 mark]

c Explain what happens to the uterus lining between days 14 and 28.

The lining of the uterus stays

_____ so it is ready

to receive _____ .

[2 marks]

> **Remember**
> A question that asks you to 'explain' requires you to link what happens to the reason it happens.

4. The graph below shows how average height of boys and girls changes with age.

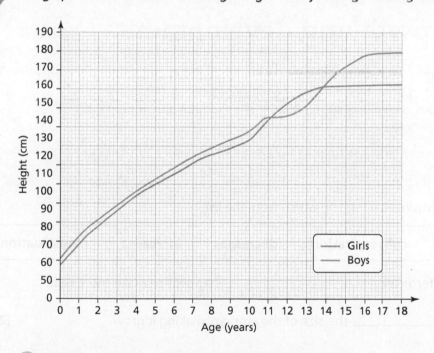

a At 15 years, the average heights of boys and girls are:

Boys: _____

Girls: _____

[2 marks]

b Between 11 and 12 years, do boys or girls grow faster, or do they both grow the same?

_____ [1 mark]

c Explain how you worked out your answer to b).

_____

_____ [2 marks]

5. Why is it important that a female's menstrual cycle should stop when she is pregnant?

Challenge

_____

_____

_____ [2 marks]

3.3 Foetal growth and development

Learning outcomes

- To describe the development of the foetus
- To explain how the foetus is protected and nourished during the gestation period
- To explain how a baby is born

1. Complete the sentences using the words from the list.

cell division	cell respiration	decreases	increases	menstruation

When an egg is fertilised, _____ happens to make more cells.

This _____ the size of the embryo, making it grow. [2 marks]

2. What is an embryo? Tick **one** box.

☐ A tiny ball of cells that grows from a fertilised egg

☐ A foetus with a beating heart

☐ An egg cell that is waiting to be fertilised

☐ A specialised cell [1 mark]

3. Draw **one** line to match each word to its correct definition.

Word	Definition
Fertilisation	Cells become specialised forming different parts of the body
Implantation	A sperm and an egg cell join together
Specialisation	Embryo becomes attached to the uterus lining

[2 marks]

4. Where does implantation happen? _____ [1 mark]

5. The table shows the length of a foetus at different times during a pregnancy.

Time in pregnancy (weeks)	9	12	16	20	24	28	32	34	40
Length of foetus (mm)	60	100	140	190	230	270	300	340	380

a Draw a line graph to show how the length of the foetus changes during the pregnancy.

[4 marks]

Show Me

b When is the fastest period of growth of the foetus? Explain your answer.

Between _____ because the

line of the graph is _____ during this period. [2 marks]

c   When is the slowest period of growth of the foetus?

Explain your answer.

Remember
A steeper slope on a graph shows a greater rate of change.

_____

_____

[2 marks]

6.   Suggest why implantation is important for a successful pregnancy.

Challenge   _____

_____

[2 marks]

7.   The placenta links the developing foetus to the mother.

Name **two** important substances that pass from the mother to the foetus.

1. _____

2. _____

[2 marks]

8.   Name **one** important substance that passes from the foetus to the mother.

_____

[1 mark]

9.   Why is a placenta described as 'an organ'?

Challenge   _____

_____

[2 marks]

10.   What is meant by 'gestation period'?

_____

[1 mark]

11.   Explain how a baby leaves its mother during birth.

_____

_____

[2 marks]

12. The table shows the energy requirements of a pregnant woman compared to a woman who is not pregnant.

Energy requirement per day (kJ)	
Pregnant woman	Non-pregnant woman
10 800	9300

a Calculate the difference in the energy requirement between a pregnant and a non-pregnant woman.

_____ [1 mark]

b Suggest **one** reason why a pregnant woman needs more energy.

_____ [1 mark]

c Why it is important that a pregnant woman has protein in her diet?

_____ [1 mark]

3.4 Drugs, disease and diet

Learning outcomes
- To describe different types of drugs
- To identify ways in which drugs, diet and diseases can affect human growth and development

1. Name **two** harmful chemicals in tobacco smoke.

1. _____

2. _____ [2 marks]

2. Describe **two** harmful effects that a pregnant woman can cause to the foetus if she smokes during pregnancy.

1. _____

2. _____ [2 marks]

3. Explain how drugs can pass from a pregnant woman to the foetus.

Show Me

Drugs diffuse from the mother's blood to the _____

in the _____ . [2 marks]

4. Cigarette smoke contains carbon monoxide. Explain how this could be harmful to a foetus.

Challenge

_____

_____

[2 marks]

> **Remember**
> If there are **two** marks for a written question, you will need to make **two** points in your answer.

5. Some scientists carried out an investigation into the effects of smoking during pregnancy on birth mass of babies born to the smoking mothers.

The table shows the scientists' results.

Number of cigarettes smoked by pregnant woman each day	Birth mass of baby (kg)
0	3.5
10	2.8
20	2.1
30	1.5
40	
50	0.5

a On the grid below, label both axes.

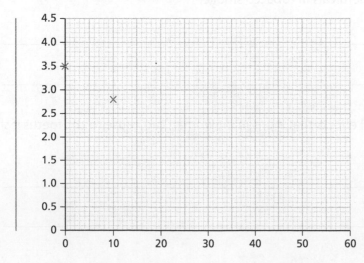

_____

[1 mark]

b Plot the points from the table as a scatter graph. The first two points have been done for you. [2 marks]

c Draw a line of best fit on the graph. [1 mark]

d Use your graph to predict the birth mass of a baby born to a mother that smokes

40 cigarettes per day. _____ [1 mark]

e Describe the pattern shown by your graph.

_____ [1 mark]

Self-assessment

Tick the column which best describes what you know and what you are able to do.

What you should know:	I don't understand this yet	I need more practice	I understand this
Male gametes are sperm cells and female gametes are egg cells			
Sperm cells and egg cells have many adaptations for their functions			
When a sperm cell nucleus and an egg cell nucleus fuse, a fertilised egg is made. This develops into an embryo, which develops into a foetus, which is born and called the baby			
Human life stages are infancy, childhood, adolescence, adulthood and old age			
Puberty is the time, during adolescence, when physical changes occur in the body			
During adolescence, people go through many physical and emotional changes.			
During puberty males and females develop secondary sexual characteristics			
The female menstrual cycle begins during puberty and includes ovulation and menstruation			
A fertilised egg develops into an embryo, which implants into the uterus lining			
At around 10 weeks, the embryo has become a foetus			
The mother's blood comes close to the foetus' blood in the placenta but they never mix			

	I can't do this yet	I need more practice	I can do this by myself
Substances diffuse between the mother's blood and the foetus' blood in the placenta			
The amniotic fluid protects the foetus			
During birth, the amniotic sac breaks, the cervix gets wider and the muscles in the uterus contract and relax to help push the baby out			
Using certain drugs during pregnancy can cause serious damage to the foetus, including reduced growth, delayed brain development and heart defects. It also increases the risk of stillbirth			
Some diseases can affect the growth and development of individuals			
You should be able to:	**I can't do this yet**	**I need more practice**	**I can do this by myself**
Interpret data from secondary sources			
Identify trends and patterns in results			
Make simple calculations			
Present results as graphs			
Test predictions with reference to evidence gained			

If you have ticked 'I don't understand this yet' or 'I can't do this yet' or mostly 'I need more practice', have another look at the relevant pages in the Student's Book. Then make sure you have completed all the questions in this Workbook chapter and the review questions in the Student's Book. If you have already completed all the questions, ask your teacher for help and suggestions on how to progress.

Teacher's comments

Test-style questions

1. The diagram shows a section through the human male reproductive system.

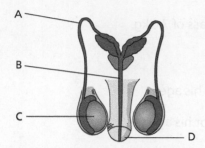

Write the letter of the label that shows the:

a urethra: _____ [1 mark]

b sperm duct: _____ [1 mark]

c structure that begins to make sperm during puberty: _____ [1 mark]

d Describe **two** other changes to males in puberty.

1. _____

2. _____ [2 marks]

e What chemicals control the changes to the body in puberty? Tick **one** box.

☐ Hormones

☐ Drugs

☐ Vitamins

☐ Enzymes [1 mark]

2. The table shows average mass of a baby boy between 0 and 6 months old.

Age (months)	0	1	2	3	4	5	6
Average mass (kg)	3.3	4.5	5.6	6.4	7.0	7.5	7.9

a Layla has a 3-month-old baby boy that has a mass of 7.0 kg.

Which statements are true? Tick **two** boxes.

☐ Layla's baby's mass is less than average for his age

☐ Layla's baby's mass is more than average for his age

☐ Layla's baby has the average mass for his age

☐ Layla's baby has the average mass for a 4-month-old boy   [2 marks]

b Between which months is a baby boy's growth the fastest? Explain your answer.

_____

_____   [2 marks]

3. Amit and Linda want to have a baby.

a Name the gametes that Linda and her husband Amit must produce for a pregnancy to occur.

Amit: _____

Linda: _____   [2 marks]

b What is the name of the process when the nuclei of gametes fuse together?

_____   [1 mark]

c On which day in Linda's menstrual cycle is she most likely to ovulate?

☐ Day 1   ☐ Day 5   ☐ Day 14   ☐ Day 27   [1 mark]

d Linda becomes pregnant. Complete the sentences using the words from the list.

| cervix | foetus | gamete | oviduct | uterus |

The embryo attaches to the lining of the _____ . Here the embryo

grows to become an unborn baby, called a _____ .   [2 marks]

e Linda is told that women should not smoke during pregnancy.

Name the organ where harmful products can pass into the blood of an unborn baby.

_____   [1 mark]

f Name **one** other harmful substance which may be passed from the mother's blood to the baby's blood.

_____   [1 mark]

g List **two** useful substances that pass from Linda's blood to the blood of the foetus.

1. _____

2. _____   [2 marks]

h Linda's friend Hassa cannot have a baby because she has blocked oviducts.

Suggest how blocked oviducts prevent a pregnancy happening.

_____

_____

_____

_____   [2 marks]

Chemistry

4.1 More on changes of state

Learning outcomes

- To use particle theory to explain the properties of solids, liquids and gases and what happens during changes in state

1. Draw **three** lines to match the particle diagram to the correct state of matter.

| liquid | gas | solid |

[2 marks]

2. The particle diagrams below show how the particle arrangement changes during a change of state.

Name the change of state.

Tick **one** box.

☐ Melting ☐ Evaporating ☐ Freezing ☐ Condensing [1 mark]

3. Complete the sentences by choosing the correct words from the list.

| freeze | look | fixed | evaporate | flow | random | slide | shape |

Solids have a definite _____ because the particles all have

_____ positions and are all touching.

Liquids _____ because the particles can _____ past each other. [4 marks]

4. When chocolate is left in the warm sun, it starts to melt.

Worked Example

Use your knowledge of particle theory to explain why.

Heat energy transferred from the Sun to the chocolate increases the vibration of the chocolate particles. ✔ This increased vibration breaks down some of the attractive forces between the chocolate particles in the solid. ✔ The solid melts to give liquid chocolate where the chocolate particles can slide past each other easily. ✔

Remember

Use your scientific knowledge to explain means to apply what you have learned in science to a new context or problem.

[3 marks]

5. Look at the diagram.

It shows a kettle of boiling water.

Remember
Compare means to describe the similarities and differences between at least two things.

Compare the water particles at point A with those at point B.

At point A water is in the _____ state. The particles are rapidly

_____ . At point B water is in the _____ state.

The particles _____ [4 marks]

6. Sometimes the pipes carrying water in houses in northern Europe split open or burst. Burst water pipes are more common in winter than in summer.

Explain why.

_____

_____ [2 marks]

7. Gases and liquids take on the shape of their container.

Challenge Bromine gas completely fills the bottle but liquid bromine just covers the bottom of the bottle.

Explain why.

_____

_____

_____ [2 marks]

Investigating changes of state

Morag and Jesse were investigating how the temperature of icy water changes as it is heated over a period of time.

At the start of the investigation they made a prediction. It is shown in the predicted graph below.

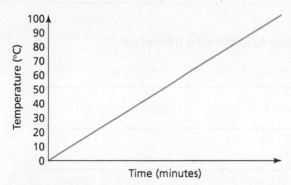

a Write down a prediction based on this graph.

_____ [1 mark]

The girls then collected some data using the following method:

- Place some crushed ice and water in a beaker

- Stir the contents of the beaker

- Measure and record the temperature using a thermometer

- Heat the beaker of ice (keep stirring)

- Record the temperature every minute until the water boils

Next the girls drew a results graph using the data they had collected.

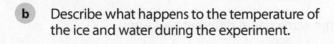

b Describe what happens to the temperature of the ice and water during the experiment.

_____

_____

_____ [4 marks]

c Compare the results graph with the predicted graph.

_____

_____ [2 marks]

d Use your knowledge of particle theory to explain the differences.

_____

_____

_____

_____ [4 marks]

4.2 Gas pressure and diffusion

Learning outcomes

- To use particle theory to explain gas pressure and diffusion
- To explain how temperature affects the rate of diffusion

1. Diffusion will occur if there is a difference in:

Tick **one** box.

☐ Mass   ☐ Pressure

☐ Temperature   ☐ Concentration   [1 mark]

2. The two particle diagrams show the same gas at different pressures.

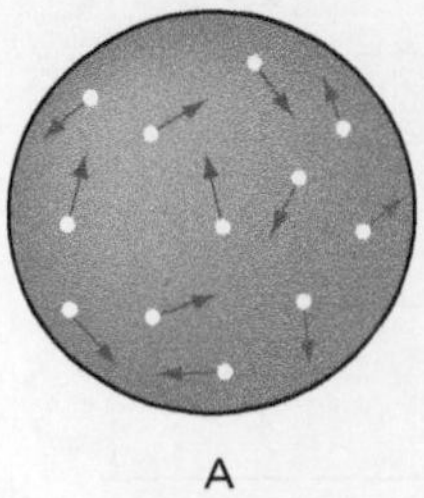

A 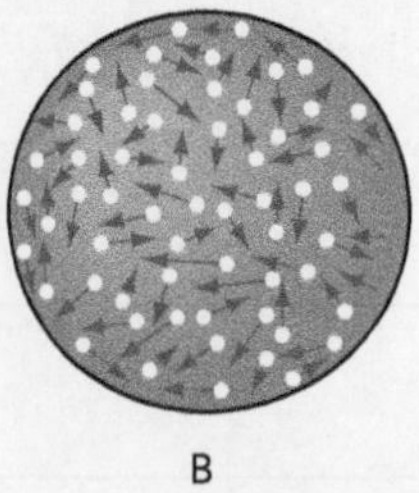   B

State which diagram shows the gas at high pressure. _____ [1 mark]

3. Which statements are true?

Tick **two** boxes.

☐ Gas pressure is a measure of particle force

☐ Gas pressure is a measure of the average particle force on the container walls

☐ Increasing the temperature causes particles to move faster

☐ Gas pressure decreases as the number of particles in the container increases [2 marks]

4. Complete the sentences by choosing words from the list below.

movement	slower	at the same speed	faster	diffusion	vibration

Diffusion occurs because of the _____ of particles in a gas or liquid.

Gas particles move _____ than liquid particles. So _____ in

liquids occurs _____ than in gases. [4 marks]

5. A crystal of lead nitrate was placed at one side of a Petri dish filled with water and a crystal of potassium iodide at the other.

After a few minutes a new coloured compound is formed between them, as shown in the diagram.

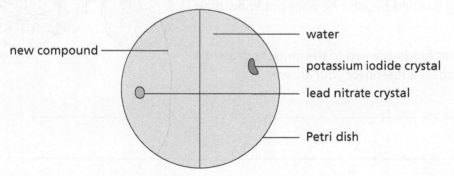

new compound

water

potassium iodide crystal

lead nitrate crystal

Petri dish

Use your scientific knowledge to explain where the new compound came from.

Both crystals dissolved _____ .

The particles _____ across the Petri

dish. When they met, a _____ took place

and a new compound was formed. [3 marks]

Remember

Use your scientific knowledge to explain means to apply what you have learned in science to a new context or problem.

6. Look at the diagram. It shows a sealed gas syringe.

Gas syringe

plunger

Rubber seal

a Describe what will happen to the pressure of the gas as the plunger is pushed in.

_____ [1 mark]

b Use your knowledge of particle theory to explain your answer to part (a)

_____

_____ [2 marks]

7. Pierre was about to go out when he notices that his bicycle has a flat tyre.

He is looking worried because he does not know how to repair it.

Describe what has happened to the amount of air in the tyre. Suggest a reason for your answer.

_____

_____ [2 marks]

8. A perfume was sprayed at the front of the classroom.

The students predicted that the whole class would be able to smell the perfume.

A minute later the students sitting near the front could smell the perfume but the students at the back could not.

Explain this observation.

_____

_____

_____ [3 marks]

9.

Challenge

Practical

When concentrated hydrochloric acid reacts with concentrated ammonia solution, a white compound is formed.

A teacher set up the following demonstration.

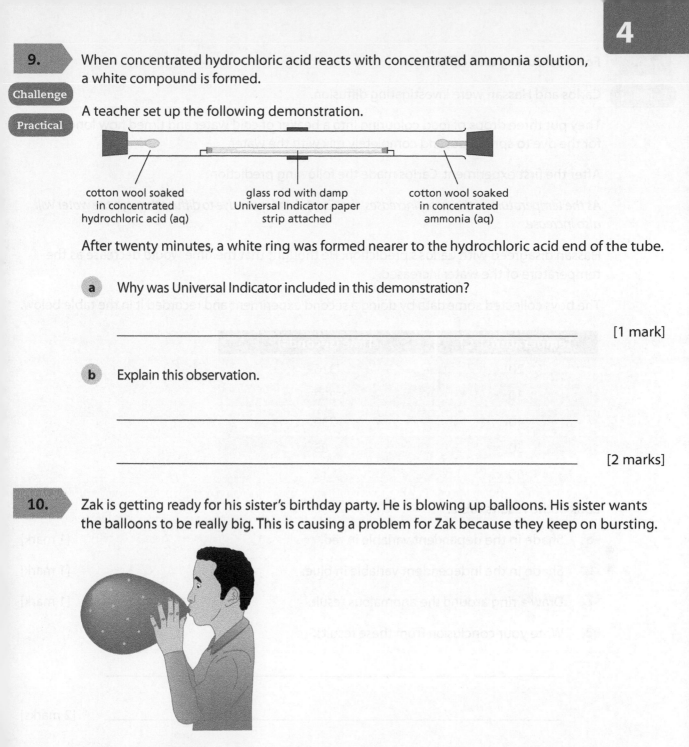

cotton wool soaked in concentrated hydrochloric acid (aq)

glass rod with damp Universal Indicator paper strip attached

cotton wool soaked in concentrated ammonia (aq)

After twenty minutes, a white ring was formed nearer to the hydrochloric acid end of the tube.

a   Why was Universal Indicator included in this demonstration?

_____   [1 mark]

b   Explain this observation.

_____

_____   [2 marks]

10.   Zak is getting ready for his sister's birthday party. He is blowing up balloons. His sister wants the balloons to be really big. This is causing a problem for Zak because they keep on bursting.

Explain why the balloons burst.

_____

_____   [2 marks]

11.

Food colouring and diffusion

Carlos and Hassan were investigating diffusion.

They put three drops of food colouring into a beaker of cold water and timed how long it took for the dye to spread out and completely mix with the water.

After the first experiment, Carlos made the following prediction.

As the temperature of the water increases, the time taken for the dye to diffuse through the water will also increase.

Hassan disagreed with Carlos's prediction. He thought that the time would decrease as the temperature of the water increased.

The boys collected some data by doing a second experiment and recorded it in the table below.

Temperature (°C)	Time (seconds)
20	316
30	222
40	230
50	88
60	12

Look at the results table.

a Shade in the dependent variable in red. [1 mark]

b Shade in the independent variable in blue. [1 mark]

c Draw a ring around the anomalous result. [1 mark]

d Write your conclusion from these results.

_____

_____ [2 marks]

e Using your answer from part (d), explain if Carlos's original prediction was correct.

_____

_____ [2 marks]

4.3 Atoms, elements and the Periodic Table

Learning outcomes
- To understand that elements are made of atoms
- To remember the symbols of the first 20 elements of the Periodic Table

1. Which statement is **not** true?

Tick **one** box.

☐ All elements are listed in the Periodic Table

☐ Each element is made from one type of atom

☐ Each element contains many different types of atoms

☐ All matter is made up of atoms

[1 mark]

2. Draw a line to match the chemical symbol to its name

| C | Ca | K | P |

| Potassium | Carbon | Phosphorus | Calcium |

[3 marks]

3. Crack the code.

Write down the element symbols to find the hidden word.

a chlorine   oxygen   carbon   potassium

_____ [2 marks]

b carbon   hydrogen   iodine   sodium

_____ [2 marks]

4. Explain why chemists find the Periodic Table so useful.

Show Me

Use words or phrases from this list to complete the answer.

| grouped together   don't have to remember them |
| randomly placed   evaluate   predict   consider |

The Periodic Table lists the names and symbols of all

known elements, so chemists _____ Elements with similar properties are

_____. This allow chemists to _____ how different elements will

react.

[3 marks]

> **Remember**
> 'Explain' means that you should say what is useful and 'why' it is useful. Where there are three marks you should make three points.

Look at the diagrams A, B and C

State which diagram represents an element.

Give a reason for your answer

_____

_____ [2 marks]

6.

Challenge

Explain why chemists use chemical symbols.

_____

_____ [2 marks]

7.

Activity focused

Use the clues provided to complete the crossword puzzle.

Across

2   The element's chemical symbol is Ar

4   The element's chemical symbol is Mg

5   About 21% of the Earth's atmosphere

6   About 78% of the Earth's atmosphere

7   Often used to make rings and jewellery

9   This gas helps clean swimming pools

10 The element's chemical symbol is Be

12 This gas can be used to fill party balloons

Down

1   This element is commonly found as soot

3   You can find all the elements listed here

8   The simplest particle in the particle theory

11 This element is used to make bridges
[10 marks]

4.4 Compounds and formulae

Learning outcomes

- To describe the difference between an element and compound
- To identify hazards and plan how to control risks when carrying out an experiment

1. Draw a line to match the key word(s) to its diagram.

| Element | Mixture of elements | Compound | A mixture of compounds | [3 marks] |

2. Complete the sentences by choosing the correct words from the list below.

| compound | nitrogen | air | oxygen | element | oxide | metal |
| nitrate | mixture |

When magnesium burns it reacts with _____ to form a new

_____ called magnesium _____ . [3 marks]

3. Name the elements present in the compound sodium chloride.

_____ [2 marks]

4. What is the formula of copper carbonate?

Tick **one** box.

☐ CuCl ☐ CaCO ☐ $CuCO_3$ ☐ HCO_2H [1 mark]

5. Josh and Adam were discussing sea water.

Show Me

Adam said that sea water was a compound but Josh disagrees. He thinks that it is a mixture.

Explain why Josh is correct.

Pure water is a _____ made of two hydrogen atoms joined to one

oxygen atom. Sea water is a mixture because it contains _____ as well as water molecules. [2 marks]

6. Mia and Ahmed heated a mixture of iron and sulfur using the equipment below.

mineral wool — clamp
— iron and sulfur
— Bunsen burner

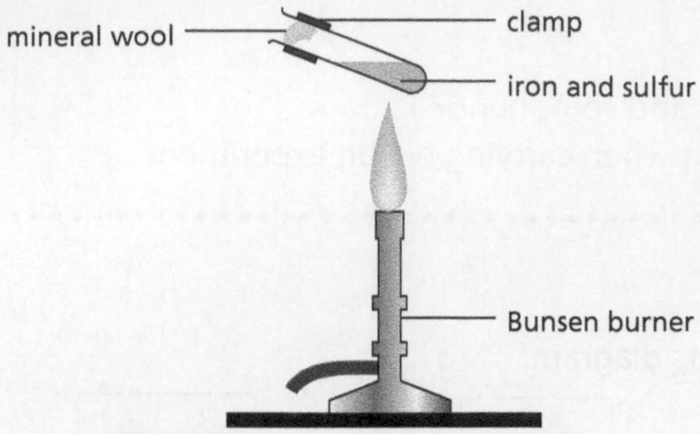

a Why did they add mineral wool to the end of the test tube?

_____

_____ [1 mark]

b What could they do to reduce the risk in this experiment?

_____

_____

_____ [2 marks]

7. Copper sulfate and diamond both form crystals.

The formula of copper sulfate is $CuSO_4.5H_2O$.

The formula of diamond is C.

Why is copper sulfate a compound and diamond an element?

_____

_____

_____

_____ [3 marks]

8. Ershad and Pramod were heating copper metal using a Bunsen burner. At the end of the experiment, they notice that the copper had gone black. Ershad thinks that the metal is covered in soot but Pramod disagrees.

Explain why Ershad is wrong.

_____

_____

_____ [3 marks]

9. **Investigating a mixture**

Two girls were investigating mixtures. They decided to test the following hypothesis:

As the temperature of the water increases, the amount of substance dissolved in the mixture also increases.

Here is their method:

- Measure out 25 cm³ water into an insulated beaker
- Measure the temperature
- Weigh out 150 g of sugar
- Add a spatula of sugar to the water in the beaker
- Stir until all the sugar has dissolved
- Add another spatula of sugar and stir
- Repeat adding sugar and stirring until no more sugar will dissolve
- Weigh the remaining sugar and work out how much has been added
- Repeat at four more temperatures

Temperature (°C)	Mass of sugar added (g)
20	50
40	60
60	73
80	88
100	120

a State **two** safety precautions the girls should take during the experiment.

_____

_____ [2 marks]

b Use graph paper to plot a graph with mass of sugar added on the *y*-axis. [2 marks]

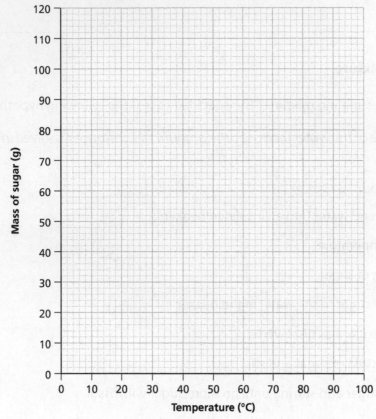

c Explain if their hypothesis is correct? Your answer must be supported by evidence.

_____

_____

_____ [2 marks]

4.5 Separating mixtures

Learning outcomes

- To explore how to separate mixtures based on their composition
- To plan a way of separating a mixture

1. Which method you would use to separate salt from a solution of salt water?

Tick **one** box.

☐ Filtration   ☐ Evaporation

☐ Sieving   ☐ Use a magnet   [1 mark]

2. Which method you would use to separate copper turnings (small flakes of copper) from a mixture of iron filings and copper turnings?

Tick **one** box.

☐ Distillation   ☐ Evaporation

☐ Filtration   ☐ Use a magnet   [1 mark]

3. Sam has been asked to separate the sand from a mixture of sand and gravel.

Practical   He uses this equipment.

Label the diagram.   [3 marks]

4. Seama has been asked to separate sand from a mixture of sand and salty water.

Practical She uses this equipment.

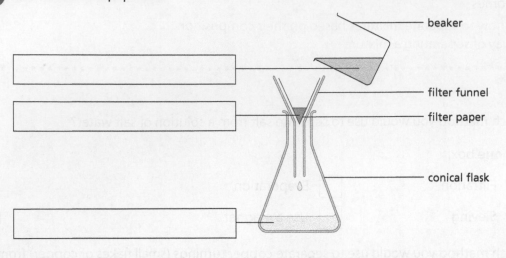

beaker

filter funnel

filter paper

conical flask

Label the diagram.   [3 marks]

5. When a mixture of salty water is separated into water and salt, distillation can be used to collect the water.

Name the **two** changes of state that occur during distillation.

Choose **two** answers from the list below.

| melting | condensation | sublimation | boiling | evaporation |

_____

_____   [1 mark]

6. Sarah wants to separate a mixture of sand and salt.

Show Me

She uses the following method.

Step 1   Mixes the sand-salt mixture with water and stirs

Step 2   Filters the mixture into a conical flask

Step 3   Pours the filtrate into an evaporating basin and heats it.

Explain why Sarah chose this method.

She mixes the sand-salt mixture to dissolve the _____. Sand is

insoluble in water so it will _____ Finally the _____

leaving the salt crystals behind.   [3 marks]

7. A mixture contains 100 g of water and 10 g of salt.

State the mass of salt left if all the water evaporates. Explain your answer.

_____

_____   [2 marks]

8. A dish contains a black and yellow powder, which you predict is a mixture of iron and sulfur.

Challenge

Practical

When a magnet is moved over the dish, black particles fly up to the magnet leaving a yellow powder behind.

State if the results support your prediction.

Explain the results of this test.

_____

_____

_____   [3 marks]

9. Separating mixtures

Activity focused

Sasha is growing crystals. She wants to grow some copper sulfate crystals but someone has contaminated the copper sulfate with chalk.

Now she needs to get a pure sample of copper sulfate before she can start growing the crystals.

Sasha plans her method. She looks in the equipment cupboard but is not sure which pieces of equipment to use.

Help Sasha decide.

For each step:

- draw a labelled diagram   [2 marks]
- explain why you have chosen the equipment.   [2 marks]

Step 1   Mix the chalk–copper sulfate mixture with water and stir.

	Explanation
	_____
	_____
	_____
	_____

Step 2   Filter the mixture

	Explanation
	_____
	_____
	_____
	_____

Step 3   Evaporate the filtrate

	Explanation
	_____
	_____
	_____
	_____

Self-assessment

Tick the column which best describes what you know and what you are able to do.

You should know:	I don't understand this yet	I need more practice	I understand this
There are forces between particles in solids, liquids and gases			
The strength of the forces determines the properties of materials			
When a material is heated, the kinetic energy of the particles increases and the motion and closeness of the particles changes. This might result in a change of state			
Gas particles are always moving and so collide with each other and the sides of their container			
The force of gas particles colliding with their container is called gas pressure			
The particles in liquids and gases spread out from where they are at a high concentration to where they are at a low concentration. This is called diffusion			
The rate of diffusion is affected by temperature and the mass of the particles			
Elements only contain one type of atom			
Each element has a symbol			
Elements are arranged in the Periodic Table			
Compounds are formed when atoms of two or more elements combine in a chemical change			
Each compound has a formula, which shows you how many of each atom are in the compound			
The properties of a compound are different from the elements it contains			

A mixture is made up of at least two different elements or compounds			
The substances in a mixture can easily be separated			
Separation methods include filtration, evaporation and distillation			
You should be able to:	**I can't do this yet**	**I need more practice**	**I can do this by myself**
Compare results with predictions			
Discuss explanations for results using scientific knowledge and understanding. Communicate these clearly to others			
Discuss and control risks to themselves and others			
Plan investigations to test ideas			
Use a range of equipment correctly			

If you have ticked 'I don't understand this yet' or 'I can't do this yet' or mostly 'I need more practice', have another look at the relevant pages in the Student's Book. Then make sure you have completed all the questions in this Workbook chapter and the review questions in the Student's Book. If you have already completed all the questions, ask your teacher for help and suggestions on how to progress.

Teacher's comments

· ·

Test-style questions

1. This question is about elements, mixtures and compounds.

a Draw **one** line to match each element with its correct chemical symbol.

Helium		B
Hydrogen		Be
Beryllium		H
Boron		He

[3 marks]

b Complete the sentence.

Choose from the words in the list.

element	**mixture**	**compound**

When iron and sulfur are mixed together and heated they form

a new _____ .

[1 mark]

c Magnesium burns in air to produce magnesium oxide.

Complete the word equation:

magnesium + oxygen → _____

[1 mark]

2. The diagram shows drawings of apparatus you could use to separate mixtures.

a Which piece of equipment would you use to:

Write down the letter.

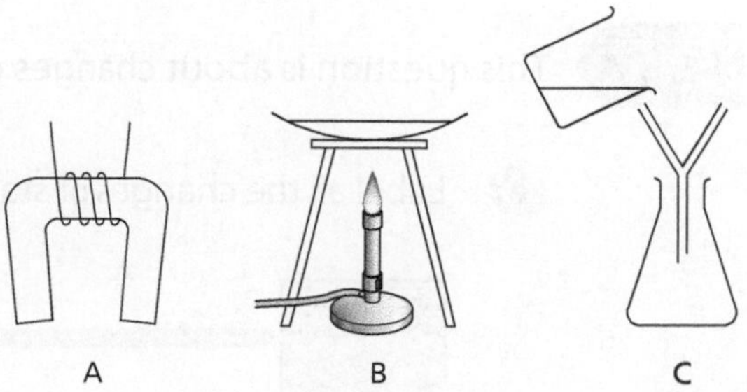

(i) Separate sand from sandy water?

_____

A   B   C

(ii) Separate iron nails from sand?   _____

(iii) Separate salt from salty water?   _____

[3 marks]

b What precautions should be taken when using the equipment labelled B?

_____

_____ [2 marks]

3. This question is about gases.

a When air is pumped into a tyre, the tyre inflates (gets bigger) as the air fills it.

Which statement about gases is correct?

Tick **two** boxes.

☐ The particles in a gas expand

☐ The particles in a gas have fixed positions

☐ Gases have no fixed shape

☐ The particles in a gas only vibrate

☐ Gases have no fixed volume [2 marks]

b Describe what will happen to the gas pressure when more air is pumped into the tyre.

_____ [1 mark]

c Explain your answer to (b).

_____

_____ [2 marks]

4. This question is about changes of state.

a Label all the changes of state shown on the diagram.

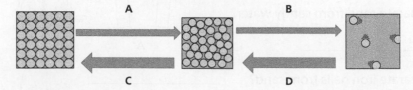

A: _____

B: _____

C: _____

D: _____   [4 marks]

b Describe what happens to the particles when a substance melts.

_____

_____   [2 marks]

c The melting point of calcium is 842 °C.

The melting point of lithium is 180.5 °C.

Explain why the melting points are different.

_____

_____

_____   [3 marks]

5.1 Metals and non-metals

Learning outcomes

• To describe and explain the differences between metals and non-metals

1. Complete the sentences using words from the list below.

more	less	solids	liquids	gases	weaker
stronger	one	two	three	states

At 20 °C most metals are _____ but non-metals will be found in

all three _____ .

Metals are usually _____ dense than non-metals.

Metals are usually _____ than non-metals which can be very brittle.   [4 marks]

2. The diagram shows the outline of the Periodic Table.

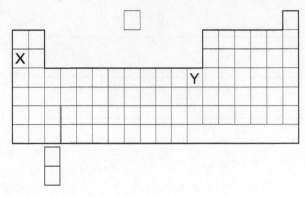

a Shade in the metals in grey.   [1 mark]

b Draw a red box around the group 1 metals.   [1 mark]

c Name elements X and Y.

_____   [2 marks]

3. This table lists the melting points of some metals and non-metals.

Worked Example

Element	Metal / non-metal	Melting point (°C)
Copper	Metal	1085
Iron	Metal	1538
Oxygen	Non-metal	−219
Chlorine	Non-metal	−102

Use the data given in the table to compare the melting points of metals and non-metals.

Copper and iron are both metals with melting points over 1000 °C. ✔

Oxygen and chlorine are both non-metals with melting points below zero. ✔

Therefore, the data shows that metals have high melting points whereas non-metals have low melting points. ✔

[3 marks]

> **Remember**
> Compare means to describe the similarities and differences of at least two things.
>
> When using negative numbers, a 'higher number' means a number with a 'lower value'. For example, −10 has a lower value than −5.

4. Density of some metal elements:

Element	Density (g/cm³)
Iron	7.87
Aluminium	2.70

Iron and aluminum metals could both be used in the construction of aircraft wings.

Use the data in the table to explain why aluminum might be a better choice than iron.

> **Remember**
> When asked to 'use the data' from the table, you must refer to it in your answer.

The density of aluminum is 2.7 g/cm³, which is about three times _____.

Therefore, the plane will be more economical to fly because it is _____.

[2 marks]

5. Jack thinks that metals such as iron and copper are harder than non-metals such as graphite.

Lily says he should do a scratch test to see if he is right.

In a scratch test objects made of a harder material will scratch objects made of a softer material.

Explain how Jack can test his prediction.

_____

_____ [2 marks]

6. Metals are often described as being malleable and ductile but many non-metals are not.

Explain the meaning of 'malleable' and 'ductile', and why non-metals do not have these properties. Your answer should include examples.

_____

_____

_____ [3 marks]

Investigating the strength of a metal wire

The diagram shows a suspension bridge.

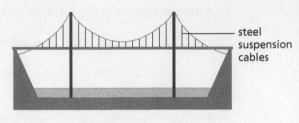

steel
suspension
cables

Nasser thinks that the cables on a suspension bridge are made from steel and not iron because steel is stronger than iron.

To test out his theory, Nasser plans an investigation using the equipment shown on the right.

He added 100 g masses until the wire broke and then recorded the maximum mass the wire held.

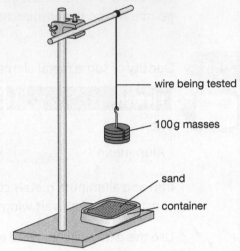

wire being tested

100 g masses

sand

container

Here are his results:

Metal	Total mass added (g)	
	1st attempt	2nd attempt
Iron	800	900
Steel	1600	1400

a Explain why Nasser put a container full of sand under the masses?

_____ [1 mark]

b State **two** variables he needed to control during the investigation.

_____

_____ [2 marks]

c Explain why Nasser tested each metal twice?

_____ [1 mark]

d Calculate the mean strength of each wire:

Iron: _____

Steel: _____

How much stronger is steel than iron?

_____

_____ [2 marks]

e Explain if the evidence supports his prediction.

_____

_____

_____ [2 marks]

f Describe two ways he could improve his method to obtain more accurate results.

_____

_____

_____ [2 marks]

5.2 Corrosion and rusting

Learning outcomes

- To describe corrosion and rusting as chemical reactions which are not useful
- To plan an investigation to test an idea

1. Draw a line to match the keyword with its meaning.

Oxidation		The deterioration of materials
Combustion		Reaction with oxygen to form an oxide
Corrosion		Reaction of iron with oxygen and water
Rusting		A burning reaction

[3 marks]

2. Complete the word equation:

sodium + _____ → sodium oxide [1 mark]

3. State which product is formed when zinc and oxygen react together.

Tick **one** box.

☐ Zinc oxygen ☐ Zinc oxate

☐ Zinc oxide ☐ Zinc oxic [1 mark]

4. Copper reacts with oxygen to form copper oxide. The mass of copper oxide formed is greater than the mass of copper used up because:

Tick **one** box.

☐ it gives off carbon dioxide

☐ oxygen has added to the copper

☐ it has burned

☐ the copper particles got bigger during the reaction [1 mark]

5. Two students were investigating rusting. They think that both water and oxygen are needed for rusting to occur.

Show Me

They set up the following experiment.

Practical

Based on the students' prediction, explain what will they expect to observe in each test tube after 5 days?

cotton wool to keep out moisture in the air

oil to stop air reaching water

air

water

water

iron screws

Rust will be observed in the first test tube because it

contains _____. _____ will be observed in test tubes 2 and 3 because they only have

_____ . [3 marks]

Remember

If questions include a prediction, read the prediction and apply it to each situation.

6. Mariam's bicycle is getting old and she is worried that it is going to go rusty.

State **two** things could she do to stop it from rusting.

_____

_____ [2 marks]

7. The girls' athletic team won a silver trophy at sports day.

When the trophy was put into the display cabinet, Mrs Patel noticed that all the other trophies looked dull and had lost their shine.

Explain why.

_____

_____ [2 marks]

8.

Challenge
Practical

Two students were investigating rusting.

After 5 days rust was only observed in test tube 1.

Explain the results in each of the test tubes

_____

_____

_____

_____ [3 marks]

water — iron screws — iron screws — grease — water — water — zinc metal wrapped around iron screws

9.

Practical

Planning an investigation

Daya is having a holiday at the seaside. When the tide goes out she notices some very rusty nails in some driftwood lying on the beach.

She thinks that the salty sea water might increase the rate of rusting of the nails.

When she gets home she decides to test out her prediction.

Daya starts by identifying the variables and setting up the experiment.

This is her method:

• She half-filled five beakers with water and labelled them 1–5

• She added different amounts salt to each beaker and stirred:

o Beaker 1 – no salt

o Beaker 2 – 1 teaspoon of salt

o Beaker 3 – 2 teaspoons of salt

o Beaker 4 – 3 teaspoons of salt

o Beaker 5 – 4 teaspoons of salt

• She weighed five nails and placed one nail in each of the five beakers

• She then observed and recorded any changes over the next five days

• Finally, she dried and re-weighed each nail at the end of the experiment.

a Complete the word equation for rusting:

iron + _____ + _____ → rust   [2 marks]

b State which variable she changed.

_____ [1 mark]

c State which variable she measured.

_____ [1 mark]

d State which variables she should have controlled.

_____

_____ [3 marks]

e Design a results table for Daya.

Write a heading for each column.

[3 marks]

f If Daya's prediction is correct, explain how would you expect the mass of the nails to change during the investigation?

_____

_____ [2 marks]

Self-assessment

Tick the column which best describes what you know and what you are able to do.

You should know:	I don't understand this yet	I need more practice	I understand this
Metals are found on the left side of the Periodic Table and non-metals on the right			
Most metals are shiny, hard, strong and dense. They have high melting and boiling points and can conduct heat and electricity			

	I can't do this yet	I need more practice	I can do this by myself
Most non-metals are dull, soft and brittle. They have low melting and boiling points and are insulators			
Many of the properties of metals and non-metals can be explained by the way their particles are arranged and the strength of the forces between them			
Metals can be damaged by chemical reactions with substances in air and water. This is called corrosion			
Corrosion makes metals weaker. It is an example of a chemical reaction that is not useful			
Rusting is the corrosion of iron. Oxygen and water must be present for iron to rust			
Rust can be prevented by a barrier that prevents air and water from reaching the surface of the iron and steel			

You should be able to:	I can't do this yet	I need more practice	I can do this by myself
Plan investigations to test ideas			
Identify important variables, choose which variables to change control and measure			
Test predictions with reference to evidence gained			
Present results as appropriate in tables and graphs			
Interpret data from secondary sources			

If you have ticked 'I don't understand this yet' or 'I can't do this yet' or mostly 'I need more practice', have another look at the relevant pages in the Student's Book. Then make sure you have completed all the questions in this Workbook chapter and the review questions in the Student's Book. If you have already completed all the questions, ask your teacher for help and suggestions on how to progress.

Teacher's comments

Test-style questions

1. This question is about chemical reactions with oxygen.

a What compound is made when iron reacts with oxygen?

Tick **one** box.

☐ Iron oxygen   ☐ Iron oxide

☐ Rust   ☐ Iron hydroxide   [1 mark]

b When carbon burns in oxygen gas, carbon dioxide gas is produced.

(i) Write the word equation for the reaction:

_____ + _____ → _____ [2 marks]

(ii) Explain why the mass of carbon appears to decrease during the reaction?

_____

_____ [2 marks]

c When magnesium is oxidised, a white solid is formed.

(i) Name the white solid. _____ [1 mark]

(ii) Predict how the mass of magnesium will change during the reaction. Give a reason for your answer.

_____

_____ [2 marks]

2. The table lists the melting points of four unknown substances:

Unknown substances	Melting point (°C)
A	2005
B	55
C	1899
D	−21

a Which unknown substances are non-metals?

_____ [2 marks]

b Name **three** properties that unknown substance A is likely to have.

_____ [3 marks]

c Unknown substance D is an element. Describe where it is likely to be found in the

Periodic Table. _____ [1 mark]

3. Priya and Gabriella are comparing different metals.

The first property they investigated was density. They measured the mass of different cubes of metal. Each cube was the same volume.

Here are their results:

Metal	Mass (g)
Copper	9.0
Iron	7.9
Aluminium	2.7
Zinc	7.1

a Name the piece of equipment they used to measure the mass of each cube.

_____ [1 mark]

b When the girls looked at the results they thought that one was wrong, so they repeated it but got the same result.

Suggest which result they repeated.

_____ [1 mark]

c Put the metals in order of increasing density.

_____ [3 marks]

d Write down two things that the girls have learned about the density of metals.

_____

_____ [2 marks]

6.1 Using word equations

Learning outcomes

- To describe reactions using word equations

1. A word equation is used to describe a chemical reaction.

   Identify the correct way of writing a word equation.

   Tick **one** box.

   ☐ reactants = products ☐ reactants → products

   ☐ products ← reactants ☐ products → reactants [1 mark]

2. Elements and compounds are made from atoms.

   Identify the element.

   Tick **one** box.

   ☐ H_2SO_4 ☐ S_8 ☐ H_2O ☐ SO_2 [1 mark]

3. The total mass of a beaker of milk and a beaker of orange juice is 125 g.

   Practical

   Show Me

   When the milk is poured into the orange juice, a cream coloured solid is formed but the mass stays the same.

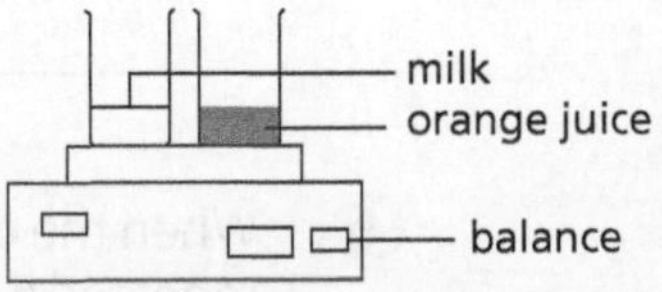

   milk
   orange juice

   balance

   Explain the observations.

   When the orange and milk was mixed a

   _____ took place. New compounds were

   formed as the _____ The mass stayed

   the same because the _____ provided all

   the atoms needed to make the products. [3 marks]

   Remember
   Questions of this type find out if you can apply your knowledge of chemical reactions to everyday materials.

4. The diagram shows the atoms present in some reactants.

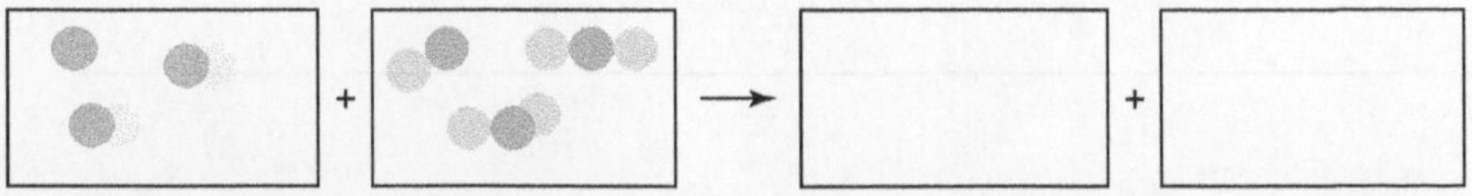

   During the reaction, the atoms rearranged to form new compounds.

   Show the atoms present in the new compounds by drawing in the last two boxes. [2 marks]

5. The diagram shows some everyday products.

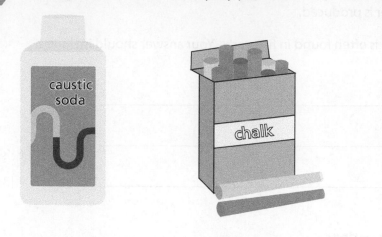

_____   _____   _____

Write the correct chemical name under each product. Use words from the list.

calcium carbonate	sodium hydroxide	calcium hydroxide
sodium hydrogen carbonate		copper oxide

[3 marks]

6. When sodium hydroxide is added to copper sulfate, you can see a blue solid (copper hydroxide) forming in colourless sodium sulfate solution.

Write a word equation for the reaction.

sodium hydroxide + copper sulfate ✔ → copper hydroxide + sodium sulfate ✔

Remember
When writing a word equation, you will find the answer in the question.

[2 marks]

7. When you add sodium hydroxide to iron(II) nitrate, you can see a green solid (iron(II) hydroxide) forming in colourless sodium nitrate solution.

Show Me

Write a word equation for the reaction.

sodium hydroxide + _____ →

iron(II) hydroxide + _____   [2 marks]

Remember
When writing a word equation, start by writing down the reactants; then draw an arrow and write down the products. Read the question carefully before you attempt the equation.

8. Farmers often put lime on fields to neutralise acidic soil.

This is the label usually found on a bag of lime.

CaO

State two things that the label tell us about lime.

_____

_____

[2 marks]

Magnesium metal burns with a bright white light. At the end of the reaction, white magnesium oxide powder is produced.

Suggest why magnesium is often found in fireworks. Your answer should include a word equation.

_____

_____

_____ [3 marks]

Investigating chemical reactions

Nuri was doing a series of experiments. He has noted down his observations in a table.

Test	Method	Observations	Conclusion
1	Add 1 drop of sodium hydroxide to iron(III) sulfate	A rust coloured solid appeared	
2	Add 1 drop of sodium hydroxide to calcium chloride	A white solid appeared	
3	Heat copper metal in a flame	The metal went black	
4	Add copper metal to hydrochloric acid	No changes observed	
5	Add magnesium metal to hydrochloric acid	Bubbles seen coming off the metal	

a Complete the table.

Write a conclusion for each experiment based on the observations. [5 marks]

b Write a word equation for each reaction. You may need to look up the names of some of the products if you cannot work them out.

Test 1:

_____ + _____ → _____ +

_____   [2 marks]

Test 2:

_____ + _____ → _____ +

_____   [2 marks]

Test 3:

_____ + _____ → _____   [2 marks]

Test 4:

_____ + _____ → _____   [2 marks]

Test 5:

_____ + _____ → _____ +

_____   [2 marks]

11. When sodium hydroxide is added to a solution of iron(III) chloride, the following can be seen:

Challenge
- a rust-coloured solid is formed
- the total mass does not change.

Explain the observations. Your answer should include a word equation.

_____

_____

_____

_____ [4 marks]

6.2 Reactions with acid

Learning outcomes

• To describe and write word equations for the formation of chloride and sulfate salts using reactions with acids
• To describe how to test a gas to see if it is hydrogen or carbon dioxide

....

1. Name the compounds produced when hydrochloric acid reacts with calcium oxide.

Tick **one** answer.

☐ calcium oxide + water   ☐ calcium sulfate + water

☐ calcium chloride + water   ☐ calcium carbonate + water   [1 mark]

2. Complete the sentences.

Chose words from the list.

| sulfate | chloride | carbonate | oxygen | hydrogen | carbon dioxide |

When magnesium metal reacts with hydrochloric acid, magnesium _____

and _____ gas are produced.

When copper carbonate reacts with sulfuric acid, copper _____ , water

and _____ gas are produced.   [4 marks]

3.

Show Me

During a chemical reaction between hydrochloric acid and sodium carbonate, Ling observed bubbles of gas being formed. She thought the gas was carbon dioxide.

Describe a test Ling could do to see if she was right.

Bubble the gas through _____ . If it changes from

_____ the gas is carbon dioxide.   [2 marks]

> **Remember**
> When describing a chemical test, you must write down what you will do and what you expect to see for a positive result.

4. When zinc metal was added to some sulfuric acid, small bubbles of gas were observed. Describe a test that would show if the gas produced was hydrogen.

_____

_____   [2 marks]

5.

Challenge

Explain what is meant by the statement:

New compounds are formed during a chemical reaction.

Give an example in your answer.

_____

_____ [3 marks]

6.

Activity focused

Practical

Making copper chloride crystals

Aki wanted to make some copper chloride crystals. He started by reacting copper oxide with hydrochloric acid. After filtering off the excess copper oxide, he put the blue copper chloride solution into an evaporating basin and evaporated off the water to obtain the crystals.

a Read the description and draw a ring around the reactants. [2 marks]

b To get the reaction to work, Aki had to heat the reactants.

Draw a labelled diagram of the equipment he should use.

[2 marks]

c Describe safety precautions Aki should take whilst heating the mixture.

_____

_____ [2 marks]

d Draw the apparatus showing how Aki should evaporate off the water.

[2 marks]

e   Read the description and underline the products.   [2 marks]

f   Write a word equation for the reaction.

_____   [2 marks]

g   Aki was so pleased with the final crystals that he decided to make some more but there was no copper oxide left. Name another chemical he could use.

_____   [1 mark]

h   During the initial reaction between the new reactants, Aki observed bubbles of gas being formed, which turned out to be carbon dioxide gas.

Write a word equation for this reaction.

_____   [2 marks]

6.3 Reactions with oxygen

Learning outcomes

- To describe and write word equations for the reaction of metals and non-metals with oxygen
- To discuss how ideas can be tested by carrying out investigations and collecting evidence

1.   Poor air quality is a global issue. Many oxides formed during the combustion of fossil fuels are pollutants.

Draw **one** line to match the oxide to the problem it can cause.

Sulfur dioxide		A poisonous gas that can kill people
Carbon monoxide		A greenhouse gas that leads to global warming
Carbon dioxide		An acidic gas which helps cause acid rain

[2 marks]

2.   When sulfur burns in oxygen, sulfur dioxide gas is formed.

Complete the word equation.

_____ + oxygen → _____   [2 marks]

3.

Practical

Show Me

This diagram shows some iron wool on a 'see-saw' balance.

Explain what will happen to the 'see-saw' when iron wool is heated. [3 marks]

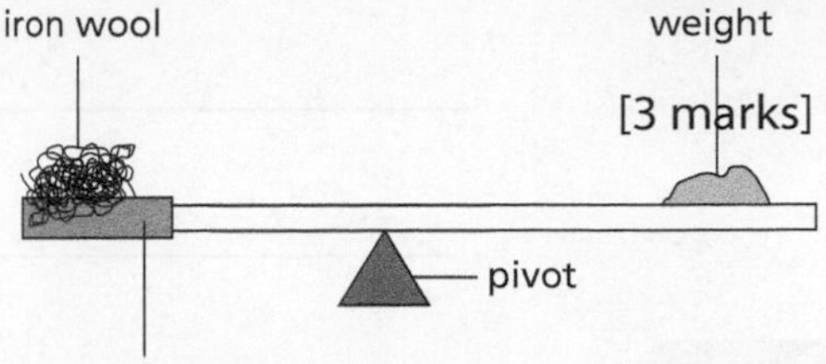

The left-hand side of the balance will

_____ This is because as the iron

wool is heated it _____ to form

iron oxide, which is _____ than air.

iron wool

weight

pivot

metal to protect ruler

Remember

'Explain' means you must say what happens and why it happens.

4.

Practical

When a candle burns in air, carbon dioxide and water are made.

The mass of a candle before burning was 15.2 g.

The mass of the candle after 5 minutes of burning was 14.4 g.

a Calculate the rate at which the candle burnt in g/min.

_____

_____ [2 marks]

b Explain why the mass decreased.

_____

_____ [2 marks]

5.

Challenge

Look at the graph. It shows how the proportion of carbon dioxide gas in the atmosphere has changed over the past 1000 years. The unit 'ppm' means parts per million.

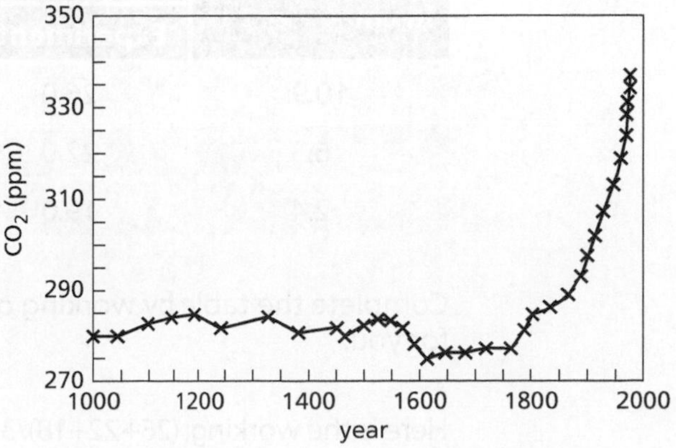

a Describe how the proportion of carbon dioxide changed between 1000 and 1500.

_____

_____ [2 marks]

b Give a reason which might explain the changes after 1800.

_____

_____ [2 marks]

Candle investigation

Mr Yung's class are investigating candles. He sets up the following equipment and asks the students to make a prediction.

Mingxia thinks the candles will burn for the same time because the beakers are the same size.

Elizaveta thinks the shortest candle will go out first because the carbon dioxide produced will sink to the bottom of the beaker and put the flame out.

a Explain why Elizaveta thinks that the carbon dioxide will sink?

_____ [1 mark]

b Explain why Mr Yung chose beakers that were the same size.

_____ [1 mark]

c Each group timed how long it took for the lighted candles to go out.

Here are the results from Mingxia's group.

Candle length in cm	Time candle burned in seconds			
	Experiment 1	**Experiment 2**	**Experiment 3**	**Mean**
10.9	26.0	22.0	18.0	22.0
6.3	22.0	23.0	17.0	
2.4	19.0	25.0	21.0	

Complete the table by working out the mean results. The first one has been done for you.

Here is the working: (26+22+18)/3 = 66/3 = 22

[2 marks]

d Write a conclusion by comparing the means of the results.

_____ [1 mark]

e   Comment on the quality of the data collected in the different experiments.

_____ [1 mark]

f   Suggest **two** ways you could improve the experiment.

_____

_____ [2 marks]

Try collecting some more data at home and see if it supports either Mingxia or Elizaveta's prediction.

Self-assessment

Tick the column which best describes what you know and what you are able to do.

You should know:	I don't understand this yet	I need more practice	I understand this
During a chemical reaction, reactants react to form new products			
A word equation separates the substances that react (on the left) and the products that are formed (on the right) with an arrow			
Simple compounds can be named using rules			
No atoms are made or lost during a chemical reaction; they are just rearranged			
Metals and bases react with acids to produce salts. Hydrochloric acid produces chloride salts and sulfuric acid produces sulfate salts			
Most metals react with acids to produce hydrogen gas			
To test if a gas is hydrogen you use a lighted splint. If a 'pop' noise is heard, hydrogen is present			
Carbonates react with acids to produce carbon dioxide			
To test if a gas is carbon dioxide you bubble the gas through limewater. If the limewater goes cloudy, carbon dioxide is present			

	I can't do this yet	I need more practice	I can do this by myself
Oxidation is a reaction with oxygen to form an oxide compound			
Combustion, or burning, is an oxidation reaction			
Some substances react with oxygen and form a solid product, while others react with oxygen and form a product which is a gas			

You should be able to:	I can't do this yet	I need more practice	I can do this by myself
Use a range of equipment correctly			
Discuss and control risks to yourself and others			
Present conclusions to others in appropriate ways			
Discuss explanations for results using scientific knowledge and understanding. Communicate these clearly to others			

If you have ticked 'I don't understand this yet' or 'I can't do this yet' or mostly 'I need more practice', have another look at the relevant pages in the Student's Book. Then make sure you have completed all the questions in this Workbook chapter and the review questions in the Student's Book. If you have already completed all the questions, ask your teacher for help and suggestions on how to progress.

Teacher's comments

Test-style questions

1. This question is about chemical reactions with acids.

a When lemon juice is added to baking soda, bubbles can be seen.

Name the gas in the bubbles.

Tick **one** box.

☐ Oxygen   ☐ Hydrogen

☐ Carbon dioxide   ☐ Sodium hydroxide   [1 mark]

b When zinc metal is added to hydrochloric acid, bubbles of gas can be observed. Omar thinks the gas is hydrogen.

Describe a test he could do to see if he was right.

_____

_____ [2 marks]

c Zinc sulfate crystals are made when zinc oxide reacts with sulfuric acid and the solution is left so that water can evaporate off.

Write a word equation for the reaction.

_____ + _____ → _____ +

_____ [2 marks]

d Describe what happens to the atoms during the reaction in (c).

_____

_____ [2 marks]

2. Mrs Brown's class were investigating how the mass of magnesium changes when it burns.

a Write a word equation for the reaction.

_____ + _____ → _____ [2 marks]

At the end of the experiment they collected the results together in a table:

Group	Mass of magnesium at start	Mass of product at end
1	0.51	0.65
2	0.52	0.72
3	0.50	0.35
4	0.51	0.75

b State what is missing from the table.

_____ [1 mark]

c Explain which group's result was unexpected. Give a reason for your answer.

_____ [1 mark]

3. This question is about poor air quality, which is a global problem.

Many pollutants are formed during the combustion of carbon-containing fuels.

a A sample of coal contains C, H and S atoms.

Which air pollutant is **not** formed by the combustion of coal?

Tick **one** box.

☐ Nitrogen dioxide   ☐ Sulfur dioxide

☐ Carbon dioxide   ☐ Water   [1 mark]

b Write a word equation for the formation of sulfur dioxide.

_____ + _____ → _____ [2 marks]

c When sulfur dioxide dissolves in rain water, the rain becomes acidic.

Describe why this might be a problem.

_____

_____ [2 marks]

d This graph shows how global temperatures have changed over the past 1000 years.

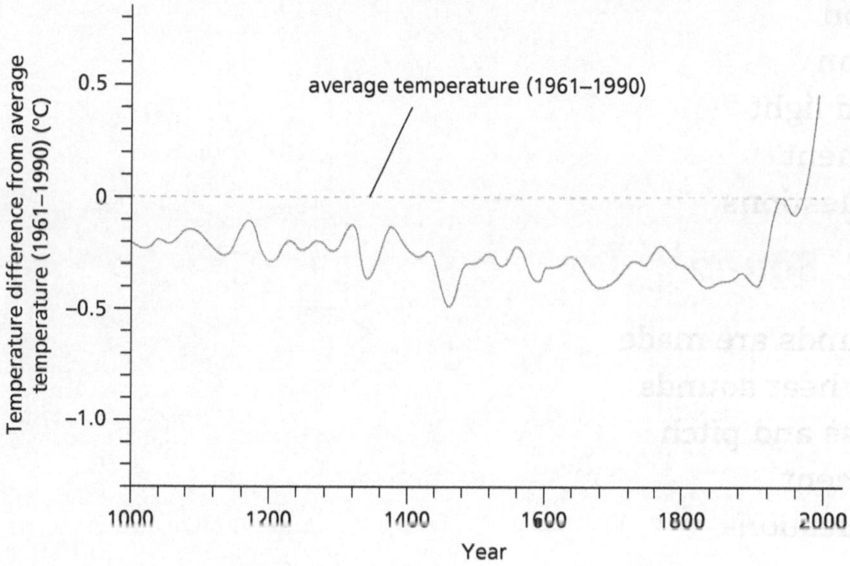

Look at the data in the graph.

Describe what the graphs shows.

_____

_____

_____ [3 marks]

e Carbon dioxide is a greenhouse gas. Explain why some scientists think that the burning of carbon fuels has caused the temperature changes seen in the graph.

Include a word equation in your answer.

_____

_____

_____

_____ [3 marks]

Physics

7.1 Light rays

Learning outcomes

- To use the idea of light travelling in straight lines to explain how images and shadows form
- To investigate how the size and shape of images and shadows may change

1. Complete the sentences using words from the list.

| object | shadow | image | straight | curved | screen |

Light only travels in _____ lines.

We can show this by shining a light on an _____ placed in front

of a _____ .

We see that a _____ forms. [4 marks]

2. Look at the diagram showing a bright light shining on an opaque object. A white projector screen is placed behind the object.

> **Remember**
> **Opaque** means no light passes through

a Draw **three** light rays to show how a pattern of light and shadow forms on the screen. [3 marks]

b What shape will the shadow form?

_____ [1 mark]

c What size will the shadow be compared to the object? _____ [1 mark]

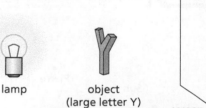

lamp   object (large letter Y)   screen

3. Anastasia starts drawing a ray diagram to show how the image is formed in her pinhole camera. Complete Anastasia's diagram by adding two more rays that show how the image is different from the real home.

> **Remember**
> A pinhole camera has a small round hole cut in one side. Light passes through the hole to form an image of an object.

Practical

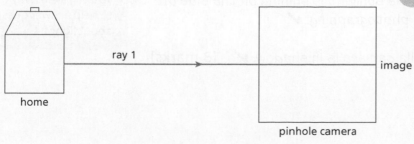

home   ray 1   image   pinhole camera

[3 marks]

Anastasia uses a pinhole camera to form an image of her home. The diagrams show four possible images. Which is the correct image? Choose the **best** image.

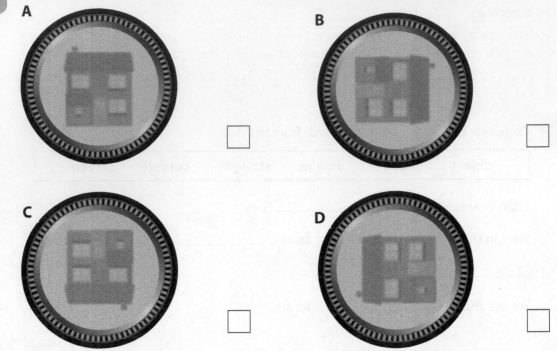

A

B

C

D

[1 mark]

Pierre and Gabriella are on holiday. They go to see a castle. They walk in different directions and each decides to take a photograph of the castle. The diagram shows where they stand as they take their photographs.

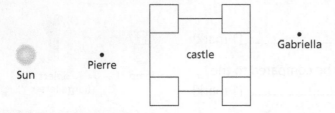

Think about light and shadows. Who do you think will take the better photograph? Explain your answer.

Pierre will take the better photograph. ✔

This is because the sunlight is shining on the side of the castle he is photographing. ✔

The side Gabriella can see is in shadow. ✔   [3 marks]

Remember
You can sketch ray diagrams to help answer questions about light.

6.

If the night is clear and there is a full Moon, sometimes you can see shadows cast by the light from the Moon.

Show Me

a Explain why we can see the Moon at night.

The light from the Sun _____ onto the dark side of the

_____ . [2 marks]

b The diagram shows the position of the Moon at four different times of the night. At which position will the Moon cast the shortest shadows? Choose the **best** answer.

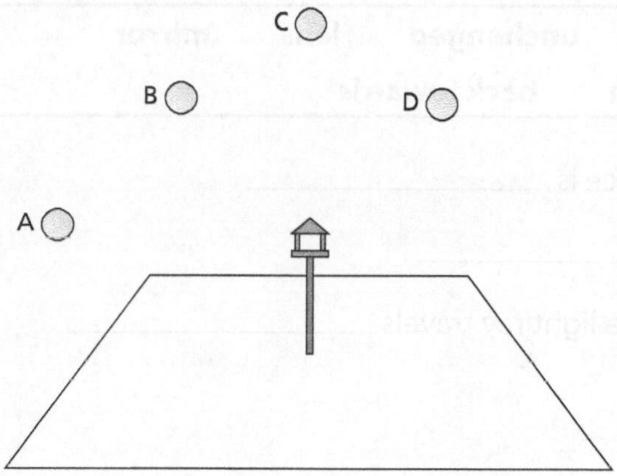

C ○

B ○   D ○

A ○

_____ [1 mark]

7.

Look at the diagram. It shows a sundial, which can be used to tell the time.

Challenge

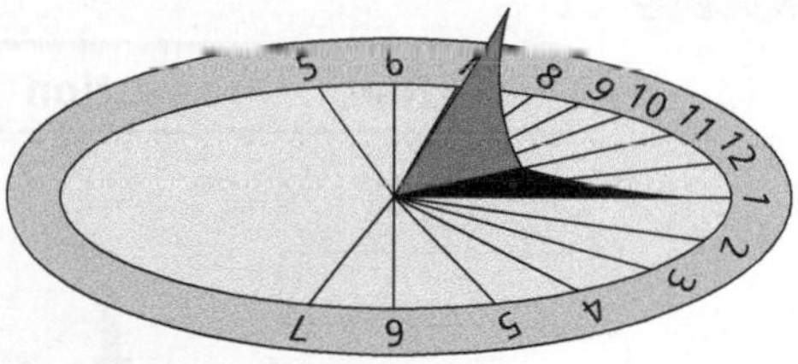

a Explain how we can tell the time of day using a sundial. (Hint: think about shadows cast by the Sun.)

_____

_____

_____

_____ [4 marks]

Practical **b** The Samrat Yantra at the Jaipur Observatory is a sundial that is 27 metres tall. The shadow from the Samrat Yantra moves at about the width of a human hand each minute. Describe an experiment you could use to work out exactly how quickly the shadow moves.

_____

_____

_____ [3 marks]

7.2 Reflection

Learning outcomes

- To describe reflection at a plane surface
- To state and use the law of reflection
- To make a periscope

1. Complete the sentences using words and phrases from the list.

refracted	reflected	unchanged	lens	mirror
	away from	back towards		

A light ray that falls on a shiny surface is _____ .

The shiny surface is called a _____ .

After it falls on the shiny surface, the light ray travels _____

_____ the source.

[3 marks]

2. Complete the labels on the diagram using words from the list.

normal	reflection	ray	refraction	incidence	object	image

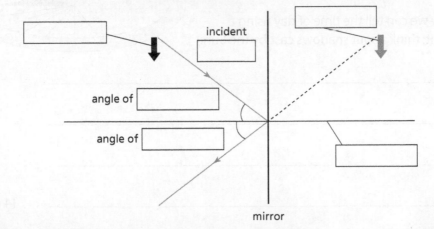

angle of ☐

angle of ☐

incident

mirror

[6 marks]

3. Describe the position of the image formed by reflection in a plane mirror and explain why it appears this way.

The position of the image appears to be

_____ the mirror.

This is because your brain thinks the light

has travelled in a _____ _____ .

> **Remember**
>
> When you **describe** something, you write what you observe. When you **explain** something, you suggest what reasons there are for what you observe.

The _____ is always as far _____ the mirror

as the object is in _____ .   [5 marks]

4.

Practical

Look at the diagram. What is the value of the angle marked X? Choose the **best** answer.

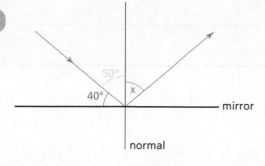

☐ A 90°   ☐ B 50°   ☐ C 40°   ☐ D 0°   [1 mark]

5.   Write down the law of reflection.

_____   [2 marks]

6.   Youssef is trying to make a sign that people can read using a mirror. He wants the sign to say 'MOUTH' when it is reflected. Write down the order of letters Youssef should put on the sign.

_____   [2 marks]

7.

Practical

The diagram shows a periscope. Two light rays are shown entering the top of the periscope.

a   Complete the ray diagram to show how both rays leave the periscope.   [4 marks]

b   State whether the image seen by the eye is the correct way up or upside down.

_____   [1 mark]

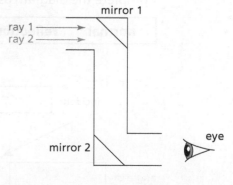

8.

Challenge

The diagram shows a device that uses mirrors to see round corners.

a   State or calculate the values of the angles A to D.

A = _____ °   C = _____ °

B = _____ °   D = _____ °

[4 marks]

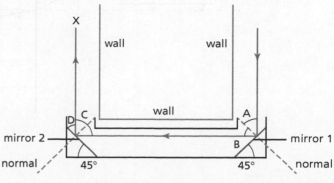

b The device is used to look at a sign that says 'HOSPITAL'. Will the image seen by the observer at point X be swapped over or still be the correct way round? Explain your answer.

_____

_____

_____ [3 marks]

7.3 Refraction

Learning outcomes

- To describe refraction
- To give examples of when refraction happens
- To do experiments to investigate how light refracts
- To draw accurately light refracting when travelling between water and air and between glass and air

1. The diagram shows a ray diagram for refraction.

Complete the diagram using the words from the list.

| normal | reflection | ray | refraction | incidence | object | image |

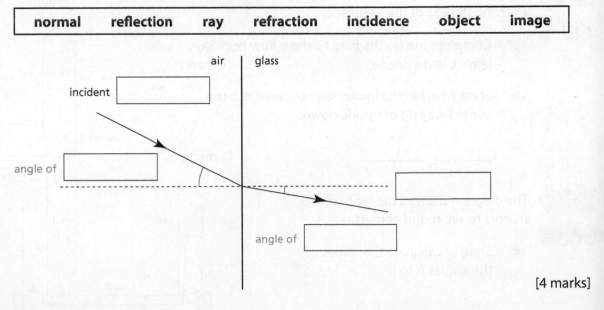

[4 marks]

2. The diagram shows arrangements of two different substances, A and B, that are placed next to each other so they touch. The table shows the possible combinations of A and B. Choose which combinations will show refraction at the boundary between A and B. Tick [✔] which rows show refraction.

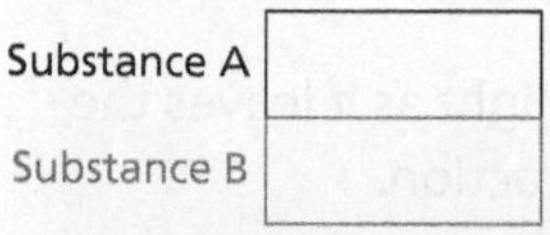

Substance A

Substance B

Substance A	Substance B	Does this combination produce refraction?
air	glass	☐
air	water	☐
glass	glass	☐
air	air	☐
water	glass	☐
glass	air	☐

[4 marks]

3. In an experiment to test refraction, state the angle of refraction if the angle of incidence is 0°.

Practical

_____ [1 mark]

4. The diagram shows the surface of a lake of water. A lamp has been placed on the bottom of the lake.

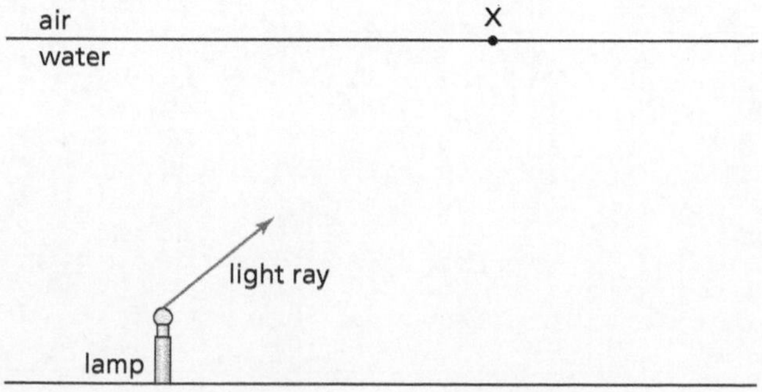

air

water   X

light ray

lamp

a Add a normal line at point X. [1 mark]

Show Me

b Describe how the direction of the light ray changes as it enters the air.

As it enters the air, the light ray _____. [1 mark]

c Complete the ray diagram to show what happens to the light as it leaves the surface of the lake. Label the angles of incidence and refraction. [3 marks]

5.

Practical

Safia is planning an experiment to investigate how different materials refract light by different amounts. This is Safia's diagram of the equipment she wants to use.

Safia needs to change the angle of incidence and measure the angle of refraction that is produced.

air

flashlight with narrow beam

substance

a Explain why it is not helpful to use an angle of incidence of 0°.

_____

_____ [2 marks]

b Suggest a piece of equipment Safia could use to measure the angles of incidence and refraction.

_____ [1 mark]

c Suggest why Safia wants to use a flashlight with a narrow beam.

_____

_____ [2 marks]

6. Look at the diagram of a glass block surrounded by air. The incident ray is shown.

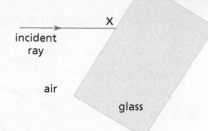

Challenge

a Think about what happens at point X. Draw a normal at X and the path of the refracted ray inside the glass block. [3 marks]

b Add the label 'Y' where your first refracted ray leaves the glass block. [1 mark]

c Think about what happens at the point you have labelled Y. Draw a normal at Y and the path of the refracted ray through the air to the right of the glass block. [3 marks]

d Add the label 'Z' to the angle of refraction between the normal at Y and the refracted ray. [1 mark]

e State the connection between the original angle of incidence at point X and the final angle of refraction Z at point Y.

_____

_____ [2 marks]

7.4 Coloured light

Learning outcomes

- To explain how white light can be separated into coloured light
- To understand the terms absorption, scattering and dispersion
- To describe examples of how different coloured lights combine
- To investigate how coloured filters work

1. Complete the sentences using words from the list.

spectrum	absorption	reflection	refraction	scattering	dispersion

A prism can be used to separate white light into a _____ .

This effect is called _____ .

When light falls on a surface and no light leaves the surface, this is called _____ .

When light falls on a rough surface and the light is reflected in many different directions,

this is called _____ . [4 marks]

2. The diagram shows white light falling on a prism.

a Complete the diagram by writing in the correct labels. [3 marks]

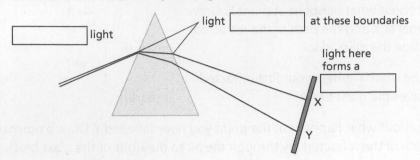

light [ ] at these boundaries

[ ] light

light here forms a [ ]

X

Y

b State the colour of the light at point X. _____ [1 mark]

c State the colour of the light at point Y. _____ [1 mark]

3. The diagram shows three large coloured lamps arranged one above the other. They shine through a thick glass block onto a screen.

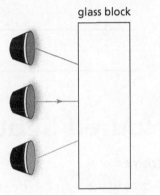

glass block

screen

a The orange light ray shown hits the block at a right-angle. Complete the ray diagram for the orange light. Label the point on the screen where the ray arrives with the letter 'X'. [2 marks]

b Add rays for the red and green lamps to show rays that also arrive at point X. [4 marks]

Show Me

c You observe the lamps from point X. Would the image you see of the lamps make them appear closer together, further apart or the same distance apart as they really are? Explain your answer.

The image makes the lamps appear

_____ .

Remember
You can work out image size by sketching a ray diagram.

From X, the images of the lamps appear to

be at a place shown by _____ .

This makes the images appear _____ the centre line. [3 marks]

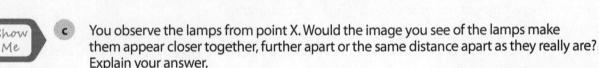

4.

Practical

Pedro planned an experiment to test how colours of light could be mixed together. Before he carried out the experiment, he predicted which colours of light would be produced when lights were combined.

The table shows his predictions.

Light 1	Light 2	Predicted colour
Red	Green	☐ Brown
Blue	Green	☐ Cyan
Red	Cyan	☐ White
Yellow	Blue	☐ Green

a Use your knowledge of coloured light to decide which of Pedro's predictions are correct. Tick [✔] each correct prediction in the table. [2 marks]

b Pedro did not know how to produce magenta light. Predict which two coloured lights could be mixed together to produce magenta.

_____

_____ [1 mark]

5.

Look at the diagram showing different surfaces. Predict what will be observed when light falls on each surface. Complete the table with your answers.

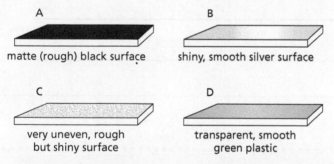

A
matte (rough) black surface

B
shiny, smooth silver surface

C
very uneven, rough but shiny surface

D
transparent, smooth green plastic

Surface	What will be observed when light falls on the surface
A	_____
B	_____
C	_____
D	_____

[4 marks]

6. Remember what you learned about energy transfers in Stage 7 Chapter 8.

Challenge
Practical

Priya designed an experiment to investigate what light does when it falls on different surfaces. She used a special red lamp to shine light on different surfaces.

Priya wrote down these results in a table.

Surface	Reflected light seen	Any other observations
Metal painted matte black	None (black)	After a few minutes the metal felt warm to the touch.
Polished metal (shiny silver)	Red light	After a few minutes there was no change in temperature of the metal.
Metal polished with a green plastic layer on top	None (black)	After a few minutes the metal temperature did not appear to change.

Explain these results.

_____

_____

_____

_____

_____

_____ [6 marks]

Self-assessment

Tick the column which best describes what you know and what you are able to do.

What you should know:	I don't understand this yet	I need more practice	I understand this
A light source gives out light			
We see things because light travels from light sources to our eyes or from light sources to objects and then to our eyes			
Light travels in straight lines			

	I can't do this yet	I need more practice	I can do this by myself
We draw the path of light with straight lines called light rays			
A shadow forms when an opaque object blocks light			
The angle of incidence = the angle of reflection			
The image in a mirror seems to be as far behind the mirror as the object is in front of it			
A periscope uses two mirrors to reflect light			
Light refracts when it travels from air into glass, clear plastic or water			
Refraction makes water look shallower from above than it really is			
A prism can split white light into a spectrum. This is dispersion			
The primary colours of light are red, blue and green			
Combining primary colours can make secondary colours and white			
Filters only allow certain colours of light to pass through them			
Filters absorb other colours of light			
Scattering takes place when light falls on a rough/uneven surface or when it hits particles in the air			

You should be able to:	**I can't do this yet**	**I need more practice**	**I can do this by myself**
Select ideas and turn them into a form that can be tested			
Plan investigations to test ideas			
Use a range of equipment correctly			
Make predictions using scientific knowledge and understanding			
Present results in tables			
Test predictions with reference to evidence gained			

Compare results with predictions			
Discuss explanations for results using scientific knowledge and understanding			

If you have ticked 'I don't understand this yet' or 'I can't do this yet' or mostly 'I need more practice', have another look at the relevant pages in the Student's Book. Then make sure you have completed all the questions in this Workbook chapter and the review questions in the Student's Book. If you have already completed all the questions, ask your teacher for help and suggestions on how to progress.

Teacher's comments

Test-style questions

1. The diagrams show four different experiments investigating light.

Name the effect each experiment shows.

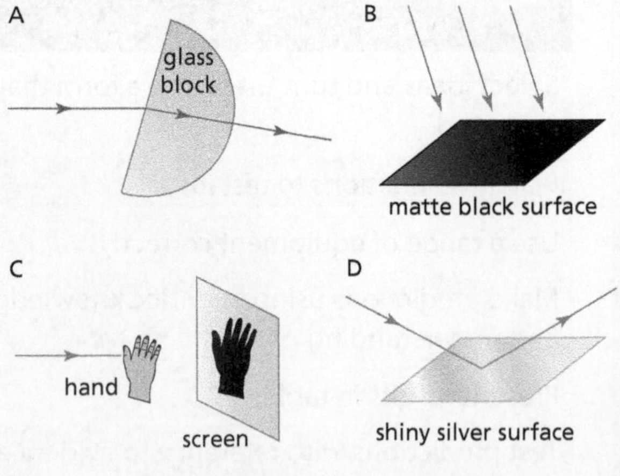

A: _____

B: _____

C: _____

D: _____ [4 marks]

A
glass block

B
matte black surface

C
hand
screen

D
shiny silver surface

2. Reflection and refraction are two different processes. Each sentence has two choices. Choose the correct choice for each sentence. Tick **one** box for each sentence.

a In reflection, the angle of incidence ☐ **equals** / ☐ **is different from** the angle of reflection.

In refraction, the angle of incidence ☐ **equals** / ☐ **is different from** the angle of refraction. [2 marks]

b In reflection in a plane mirror, the image size ☐ **is the same as** / ☐ **is different from** the object size.

In refraction, the image size ☐ **is the same as** / ☐ **is different from** the object size. [2 marks]

c Reflected words appear ☐ **the right way round** / ☐ **back to front**.

Refracted words appear ☐ **the right way round** / ☐ **back to front**. [2 marks]

3. White light passes through a red filter before arriving at a plane mirror. Then the reflected light passes through a blue filter.

State which colour of light can be seen at each of the points A to D described in the table. Choose the **best** description of the colour from the list.

| white | red | orange | yellow | green | blue | black (no light passes) |

Point	Colour of light
A: before red filter	_____
B: after red filter but before mirror	_____
C: after reflection but before blue filter	_____
D: after blue filter	_____

[4 marks]

4.

Some devices use prisms to direct light rays to a person's eyes. The diagram shows how these prisms can be arranged.

a Complete the ray diagram for one side of the device to show how the incident ray is changed so that it arrives at the eye. [3 marks]

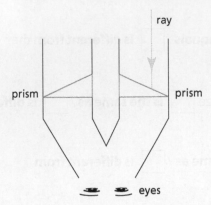

b Describe how a prism can be used to produce coloured light.

_____

_____ [3 marks]

c State the name of this effect. Choose the **best** word from the list.

dispersion	scattering	absorption

_____ [1 mark]

d Devices that use prisms can produce images that have coloured edges (fringes). Suggest a reason for this.

_____

_____ [2 marks]

e One way of reducing these fringes is to place a coloured coating over the front glass of the device. State the colour of the filter you would use to **remove** red colour fringes.

_____ [1 mark]

8.1 How sounds are made

Learning outcomes

- To describe how vibrations can produce sound
- To investigate how sounds are made
- To describe how sounds require a medium through which to travel

- -

1. Complete the sentences using words from the list.

wave	**force**	**empty**	**energy**	**vibrates**	**burns**

Sound is produced when an object or a substance _____ .

We represent how sound travels as a _____ .

Sound cannot travel through _____ space.   [3 marks]

2. Look at the diagram that models a sound wave. Complete the sentences using the words from the list.

→ direction of travel of wave

each particle vibrates

transverse	**longitudinal**	**parallel**	**perpendicular**

Sound waves are _____ waves.

This means the air particles vibrate in a direction _____ to the direction in which the wave travels.   [2 marks]

3. Yuri hits a metal bar with a small metal hammer. He places one ear against the bar and the other ear is open to the air.

Choose the **best** sentence to complete the explanation.

Tick **one** box only.

☐ Yuri hears the sound through the air first, because sound travels faster in air than in metal.

☐ Yuri hears both sounds at the same time, because sound travels at the same speed in air and metal.

☐ Yuri hears the sound through the metal first, because sound travels faster through metal than in air.

☐ Yuri only hears the sound through the air, because sound will not travel through metal.

[1 mark]

117

4. The diagrams show some materials.

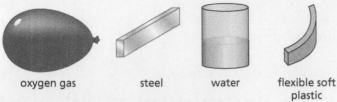

oxygen gas   steel   water   flexible soft plastic

a Choose the material that would be the best medium for sound.

_____ [1 mark]

b Explain why you chose that material.

Steel is a solid metal. Solids have a fixed structure, gases and liquids do not. ✔ Sounds are made by vibrations and the fixed structure of solids vibrates more easily. ✔   [2 marks]

> **Remember**
> When you need to write an explanation, use the number of marks to guide how much you need to write. Two marks suggests that you need two main points in your answer.

5. Carlos wants to investigate which objects can produce sounds. He has been given a diagram of the apparatus.

Practical

He has also been given some instructions about the steps he needs to take in planning his investigation. He has written some notes about the things he can choose between.

Help Carlos by matching each of his notes to the instruction. Write the **letter** of **each** statement in the correct place. The first one has been done for you.

Instructions

1 Decide which variable to change.   B

2 Decide which variable will be measured as it changes.   ☐

3 Decide which variables to keep constant (the same).   ☐

4 Note any variables that could affect the results but which you cannot control.   ☐

Carlos's notes

A Temperature and distance from microphone.

B Choice of object.

C Whether sound is detected.

D Background noise (from other people and other rooms).   [3 marks]

6.

Practical

Jamila has written down a number of ideas for investigations about sound. Choose the **best** idea from the list. Tick **one**.

☐ Do objects produce sounds underwater?

☐ The size of a musical instrument affects whether or not I can hear it.

☐ The speed of the wind affects which sounds I can hear outside.

☐ Whether or not a sound travels through an object depends on the material the object is made from.

[1 mark]

Remember
- A good investigation description should include:
- what is being changed
- what is being measured
- a clear hypothesis (an idea about how the variable being changed affects what is being measured)
- an idea that is possible to test easily.

7.

Show Me

The diagram shows a metal ruler attached to a table so that half of the ruler sticks out over the edge. Lily finds she can make a sound by bending the ruler at the end and releasing it.

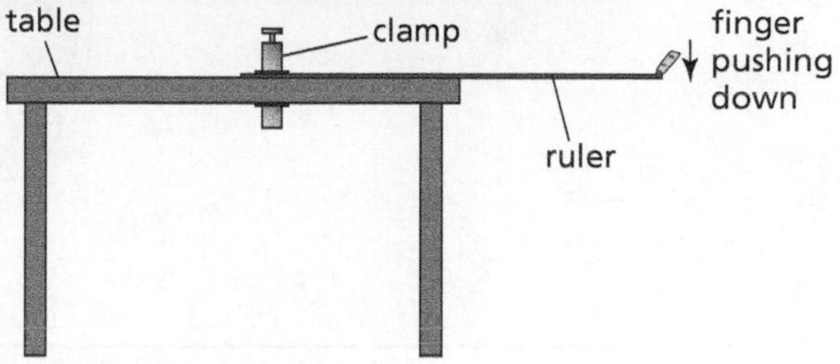

Describe the steps in the process that cause a sound to be made.

The force pushing on the ruler causes the _____

The moving ruler causes the air molecules to _____

The movement of the air molecules _____ [3 marks]

8.

Challenge

Look at the diagram. An electric circuit including a buzzer is set up inside a glass tank.

Mia switches the circuit on so the buzzer makes a loud noise. She then switches on a pump that gradually pumps the air out of the tank.

Mia observes that the sound gets quieter until she cannot hear the sound at all.

Explain Mia's observations.

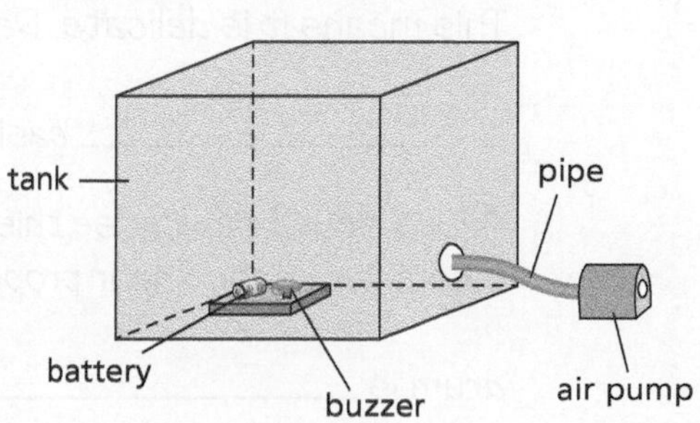

_____

_____

_____

_____ [4 marks]

8.2 How we hear sounds

Learning outcomes

- To identify the different parts of a human ear
- To describe how we hear sounds
- To describe how humans hear a limited range of sounds that changes as we get older
- To describe how we can protect our hearing from damage
- To measure the speed of sound

1. Look at the diagram of a human ear. Complete the diagram by writing in the labels. Use words from the list below. The first label has been completed for you.

nerve

| ear canal | ear drum | ossicles | nerve |

[3 marks]

2. Explain how sounds can damage our ears.

Show Me

A loud sound causes the ear drum to vibrate _____ _____ .

Our ear drum is a piece of skin tissue that is _____ _____ .

This means it is delicate. 'Delicate' means that

it _____ easily.

A loud sound can cause this to happen, which means we cannot hear properly until the ear

drum is _____ . [4 marks]

Remember

When you need to write a long explanation, plan your answer in a logical order. Write one or two short sentences for each stage of your explanation.

3. Choose which of these activities should only be carried out if we wear ear protection. Tick **three** boxes to show the **best** choices.

☐ **A** Using an electric drill on a wall.   ☐ **B** Playing football.

☐ **C** Using a mechanical drill to dig up a road.   ☐ **D** Going to a rock concert.

☐ **E** Driving a car.

[3 marks]

4. Choose the correct estimate of the speed of sound in air. Tick **one** only.

☐ A 3.4 m/s   ☐ B 340 m/s

☐ C 3.4 km/s   ☐ D 340 km/s   [1 mark]

5. Describe what happens when we hear an echo.

_____

_____   [3 marks]

6. Angelique investigates the speed of sound in different materials. The table shows her results.

Practical

Material	Speed of sound (m/s)
air (gas)	340
oxygen (gas)	3300
water (liquid)	1500
steel (solid)	6100
diamond (solid)	12 000

a One of Angelique's results is wrong. Which result is the wrong result?

_____ [1 mark]

b Ignore the wrong result. Can you see a pattern in the results that remain? Describe this pattern.

_____

_____

_____   [3 marks]

7. Safia investigates the speed of sound in steel. She calculates that it is 6100 m/s.

Worked Example

a The speed of sound in air is 340 m/s. How many more times faster does sound travel in steel than in air?

Number of times faster = speed of sound in steel / speed of sound in air

= 6100 / 340   ✔

= 18 times (accept 17.9, 17.94) ✔   [2 marks]

b The speed of sound in sea water is 1500 m/s. How many more times faster does sound travel in sea water than in air?

_____

_____ [2 marks]

8.

Practical

Challenge

Engineers testing steel rails can use a microphone to record sounds of an approaching train. The microphone records the sound of an approaching train for many seconds before the engineers can hear it themselves. The results are shown on an oscilloscope screen.

Use your knowledge of the speed of sound to explain what the engineers observe.

_____

_____

_____

_____ [3 marks]

9.

Challenge

Bats use sound to steer around obstacles in the dark. The bats produce sounds to bounce off objects. The diagram shows the face of a horseshoe bat.

Suggest why this bat's face is shaped this way.

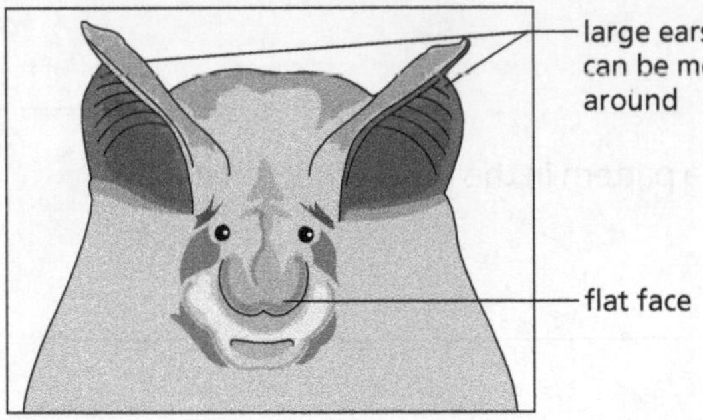

- large ears that can be moved around
- flat face

_____

_____

_____

_____ [3 marks]

8.3 Loudness and pitch

Learning outcomes

- To identify the different variables we can change and measure for waves
- To relate these variables to the sounds we can make or hear

1. The diagram shows how a wave can be represented on an oscilloscope screen.

Complete the diagram using the words from the list.

wavelength	frequency	amplitude	period

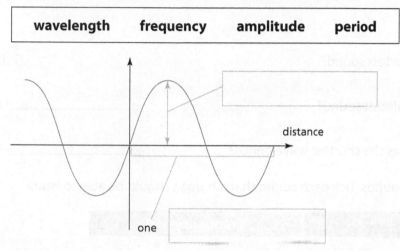

distance

one

[2 marks]

2. Look at the table. The key words 1 to 5 match up with the descriptions A to E.

For each key word, choose the **best** description. Write in a description letter next to each key word. The first one has been done for you.

Key word	Description
1 Amplitude C	A The time taken for one complete wave to pass
2 Period	B The number of waves per second
3 Wavelength	C The maximum height of a wave
4 Frequency	D The unit of frequency
5 Hertz	E The length of one complete wave

[4 marks]

3. Look at the diagrams of sound waves shown on an oscilloscope screen. They are all to the same scale. Choose the **best** diagram to answer each of the following questions. The first question has been answered for you.

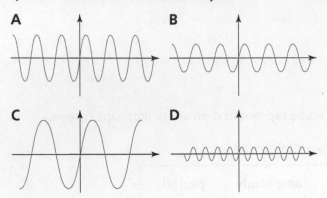

A

B

C

D

a Which is the loudest sound? *C* [1 mark]

b Which is the quietest sound? _____ [1 mark]

c Which sound has the shortest wavelength? _____ [1 mark]

4. Look at the table of sounds. Tick each sound that a human should be able to hear.

Sound	Frequency (Hz)
☐ Musical note A	440
☐ The call of a bat	45 000
☐ The song of a blue whale	10
☐ Highest note on a piano	4186
☐ Thump of a bass drum	70

[3 marks]

5. The graph shows the results of a hearing test for Ahmed. The vertical axis shows the minimum loudness Ahmed can hear for a particular frequency of sound.

Practical

If the sound is quieter than this value, Ahmed cannot hear it.

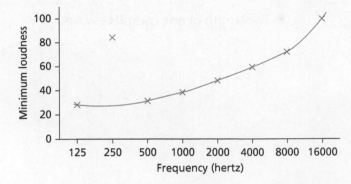

a Describe any pattern you can see in the results. Write your answer in the style of a conclusion.

The graph shows a line that _____ .

As the frequency _____ . [2 marks]

b Which result does not fit this pattern?

_____ [1 mark]

c Ahmed is 68 years old. Suggest how the graph would change for a healthy 18-year-old person.

_____

_____ [2 marks]

6. Anastasia is pregnant and has been asked to go to the hospital for an ultrasound scan. Anastasia does not know what ultrasound is. She is worried about the test. Write a short explanation of ultrasound for her. Make sure you include an explanation of why the test will not damage her baby.

_____

_____

_____ [3 marks]

7. Aiko is a sound technician for a rock band.

Challenge The band like to turn up the sound very high and play lots of electric guitar music.

Aiko measured the loudness of one of the concerts. The concert lasted two hours and the loudness changed depending on the song that was being played.

Look at the graph of loudness against time.

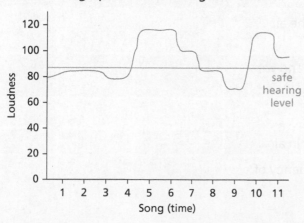

a The horizontal line shows the maximum loudness that is completely safe for humans. Was this concert completely safe for everyone that went to it? Explain your answer.

_____

_____ [2 marks]

b Suggest **two** things that Aiko could do to protect her hearing.

1. _____

2. _____ [2 marks]

c Explain to Aiko why protecting her hearing is important.

_____

_____

_____ [3 marks]

Self-assessment

Tick the column which best describes what you know and what you are able to do.

What you should know:	I don't understand this yet	I need more practice	I understand this
Vibrations produce sound			
Sound travels as a longitudinal wave			
Sound travels through the air by making the air particles vibrate			
Vibrating air particles enter our ear canals and we detect this as sound			
Echoes are sound reflections			
Sound travels at a speed of about 340 m/s in air			
The pitch of a sound depends on the frequency of the sound wave			

	I can't do this yet	I need more practice	I can do this by myself
The volume of a sound depends on the amplitude of the sound wave			
The human hearing range is from 20 Hz to 20 000 Hz			
Other animals may have different hearing ranges			
Ultrasound means sound waves which are too high for humans to hear			
You should be able to:	**I can't do this yet**	**I need more practice**	**I can do this by myself**
Select ideas for investigations and turn them into a form that can be tested			
Identify important variables and choose which variables to change, control and measure			
Identify trends and patterns in results (correlations)			
Identify anomalous results			
Understand how an oscilloscope is used to measure sounds			
Present results as appropriate in tables and graphs			
Discuss explanations for results using scientific knowledge and understanding. Communicate these clearly to others			
Present conclusions to others in appropriate ways			

If you have ticked 'I don't understand this yet' or 'I can't do this yet' or mostly 'I need more practice', have another look at the relevant pages in the Student's Book. Then make sure you have completed all the questions in this Workbook chapter and the review questions in the Student's Book. If you have already completed all the questions, ask your teacher for help and suggestions on how to progress.

Teacher's comments

Test-style questions

1. The diagrams show oscilloscope measurements of sounds made by different animals. Match the sound to the animals in the list. The first one has been done for you.

bat	blue whale	starling

A human

B _____

C _____

D _____ [3 marks]

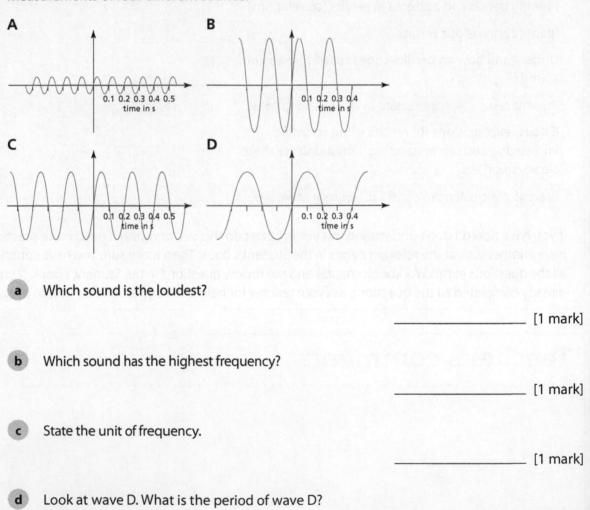

A frequency = 150 Hz

B frequency = 4000 Hz

C frequency = 45 000 Hz

D frequency = 50 Hz

2. The diagrams show oscilloscope measurements of four different sounds.

a Which sound is the loudest?

_____ [1 mark]

b Which sound has the highest frequency?

_____ [1 mark]

c State the unit of frequency.

_____ [1 mark]

d Look at wave D. What is the period of wave D?

_____ [1 mark]

3. The diagram shows a kettle drum.

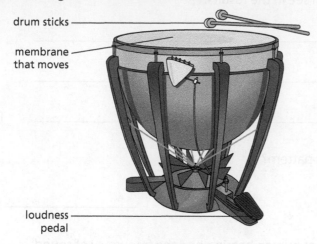

drum sticks

membrane
that moves

loudness
pedal

Write short explanations of:

a How the drum produces a sound.

_____

_____

_____ [3 marks]

b How a person hears the sound the drum produces.

_____

_____

_____ [3 marks]

4. Doctor Strange was visited by a family. Some of them were having problems hearing some sounds.

The doctor decided to test each family member's hearing by measuring the highest frequency sound they could hear. The loudness of each sound was kept the same. The results are shown in this table.

Name	Age	Highest frequency that can be heard (Hz)
Gabriella	74	9000
Carlos	50	16 500
Safia	44	18 000
Lily	24	8000
Ahmed	17	20 000

a   Describe any pattern you can see in the results.

_____

_____

_____   [2 marks]

b   Which result does not fit this pattern?

_____ [1 mark]

c   Think about what you know of human hearing and the frequency of sound.
Write a conclusion based on the results.

_____

_____

_____   [2 marks]

d   Imagine you are Doctor Strange. Who would you recommend have further tests to
check whether they might have a disease or other problem affecting their hearing?

_____   [1 mark]

9.1 Measuring distance and time

Learning outcomes

- To select suitable measuring apparatus
- To measure the distance an object travels accurately
- To measure the time for an object to travel between two places accurately

1. Complete the sentences using words from the list.

Practical

| speed | force | distance | time | mass | weight |

We can use a ruler or measuring tape to measure _____ .

We can use a stop clock or light gates to measure _____ .

We need to measure both these things to calculate an object's _____ .

[3 marks]

2. Look at the diagram of four lengths we can measure. Which measuring device is the most accurate for measuring each length? Choose the **best** device from the list.

- 15 cm ruler marked in millimetres
- metre rule marked in millimetres
- 10 m measuring tape marked in centimetres
- 100 m measuring tape marked in centimetres

A

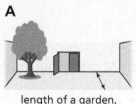

length of a garden, about 70 m

B

width of a box, about 10 cm

C

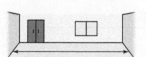

width of the room, about 6 m

D

length of a person's stride, about 80 cm

A: _____   B: _____

C: _____   D: _____

[4 marks]

Practical

Worked Example

Hassan wants to measure accurately how far he can jump. He chooses a 30 cm ruler marked in millimetres.

a Explain to Hassan why this is not the best choice of measuring device.

Hassan can probably jump about 2 m. He would need to place the ruler end-over-end about 7 times to measure the distance. ✔ This makes it likely he would make errors. ✔

[2 marks]

b Suggest to Hassan a better choice of measuring device. Explain your answer.

Show Me

A measuring _____ marked

in _____ . This will be more

accurate because it only needs to be placed

_____ .

[2 marks]

> **Remember**
> You also need to match the accuracy of the measuring device to the quantity being measured. Here, measuring to the nearest mm is probably not possible, so a device that measures to the nearest 1 cm or 0.5 cm is enough.

4.

Practical

Angelique uses a watch to record how long it takes to run 400 m. The first part of the diagram (**A**) shows the face of the watch.

A

B

12:55:73.ᴍ

a Would this watch give Angelique an accurate measurement? Explain your answer.

_____

_____

_____

[3 marks]

b The second part of the diagram (**B**) shows the face of a stop watch that Aiko uses to make the same measurement. How accurate is this watch?

_____

_____

[2 marks]

5.

Practical

The diagram shows an experiment that Oliver has set up to measure the **time** it takes for a toy car to roll down a slope at different angles.

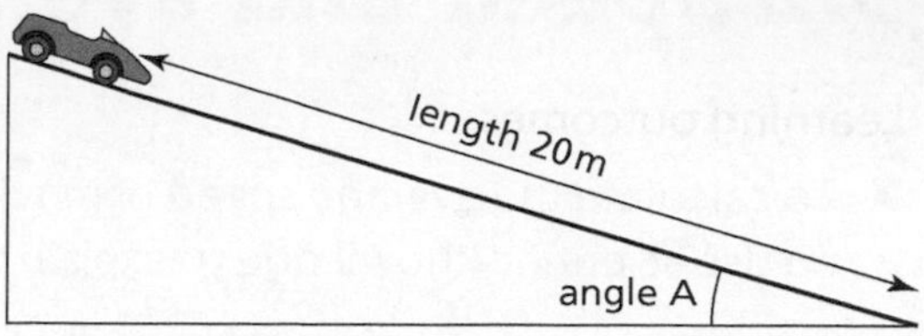

Describe the best type of device that Oliver could use for measuring the time. Include a short description of how the device works.

length 20m

angle A

_____

_____

_____ [3 marks]

6. Write down a definition of the word 'accurate'.

_____

_____ [2 marks]

7. Look at Oliver's experiment in question 5. The table shows the results of Oliver's experiment.

Challenge

Practical

Angle of slope in degrees	Time taken in seconds	Speed
20	2.0	
30	1.7	
40	1.5	
50	0.8	

a Suggest **one** way in which Oliver could improve the accuracy of his experiment.

_____ [1 mark]

b Time is the variable that Oliver measures. Which other variable does Oliver need

to measure to calculate the car's speed? _____ [1 mark]

c There are other variables that might change and affect the results. In science, we need to know about and **control** these variables to stop them affecting the results.

Suggest how Oliver could control **two** of these variables in his experiment.

_____

_____ [2 marks]

9.2 Speed and average speed

Learning outcomes

- To calculate the average speed of moving objects
- To use scientific knowledge to explain the results of experiments

1. Complete the sentences using words from the list.

speed	time	distance	metres	kilograms	hour	second	day

Speed describes how much _____ an object travels in a

given _____ .

The unit of speed is _____ per _____ . [4 marks]

2. The equation that connects average speed, distance travelled and time taken can be written in this form:

$$A = \frac{B}{C}$$

a Write down the names of the variables shown as A, B and C. Use the terms from the following list.

average speed	total distance travelled	total time taken

A = _____

B = _____

C = _____ [3 marks]

Show Me

b Explain why it is important that the total distance travelled and total time taken are used.

The totals are needed because

_____

_____ [1 mark]

3. Jamila runs 20 m in 5 s. What is her average speed? Choose the **best** answer from the list.

☐ 20 m/s   ☐ 5 m/s

☐ 4 m/s   ☐ 2 m/s [1 mark]

4. Chen cycles at 20 m/s for 5 s. What is the distance he travels? Choose the **best** answer from the list.

☐ 20 m   ☐ 4 m

☐ 50 m   ☐ 100 m   [1 mark]

5. "A 100 m sprinter can run at about 10 m/s."

Show Me

a Explain what this statement means.

A sprinter runs a race that is _____ long.

They can travel a distance of about 10 _____ in

each _____ .   [3 marks]

b Describe how a sprinter's speed changes during a race.

_____

_____   [2 marks]

c If the sprinter's **average** speed is about 10 m/s, what can you say about the **maximum** speed the sprinter reaches?

_____   [1 mark]

6. Pierre sets up an experiment to measure the time it takes a ball to fall to the floor from different heights. Look at Pierre's diagram of the equipment.

Practical

Pierre changes the height each time but always uses the same ball.

a Why does Pierre use the same ball each time?

_____

[1 mark]

b Suggest what the measuring devices labelled 'LG' are.

_____   [1 mark]

These are the results Pierre finds.

Height (m)	Time taken (s)	Average speed (m/s)
10	1.4	7.1
20	2.0	_____ [1]
5	1.0	_____ [1]

Worked Example

The first row shows the average speed for the first height. This is calculated by using the speed equation:

average speed = total distance moved
                   total time taken

$$= \frac{10 \text{ m}}{1.4 \text{ s}}$$

$$= 7.1 \text{ m/s}$$

c Calculate the values of speed for the second and third rows. Complete the table with these values.

Show Me

d Describe the pattern in these results.

Increasing the height makes the ball's average speed _____ [1 mark]

7. Migrating birds travel long distances. The world record for a non-stop flight by a bird is shown on the map. The bird is known as a bar-tailed godwit. It flew at an average speed of 50 km/h.

a One day contains 24 hours. How many kilometres will the godwit travel in 1 day?

_____

_____

[2 marks]

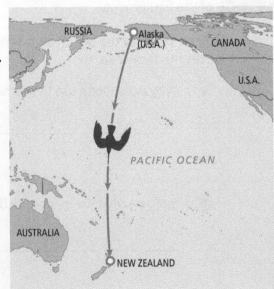

b The non-stop flight lasted nearly 10 days. One of the distances in the list is the actual distance measured. Choose the actual distance the godwit travelled.

☐ 1170 km ☐ 3024 km

☐ 8670 km ☐ 11 680 km [2 marks]

8. The average speed of a Formula 1 racing car around the circuit at Yas Marina in Abu Dhabi is 200 km/h.

Challenge

a How many kilometres does the F1 car travel in 1 minute at this speed? Show your working.

_____

_____ [2 marks]

Light gates are used to measure the speed of the F1 car along the main straight section of track. The record speed measured is 360 km/h.

b Explain why this measured speed is different from the average speed.

_____

_____ [2 marks]

c Calculate this measured speed in metres per second, m/s. Show your working.

_____

_____

_____

_____ [4 marks]

Remember what you learned in Stage 7 Chapter 8 about energy transfers.

d F1 cars need special brakes to slow them down. The brakes must work at very high temperatures (over 1000 °C). Use your knowledge of energy transfers and the conservation of energy to explain why the brakes must work at high temperatures.

_____

_____

_____ [3 marks]

9.3 Distance–time graphs

Learning outcomes

- To use graphs to show how an object moves
- To understand a distance–time graph for a moving object
- To draw a distance–time graph for a moving object
- To use a distance–time graph to calculate speed

1. The diagram shows a distance–time graph. Complete the labels using the words from the list.

| time | speed | distance |

[3 marks]

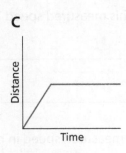

gradient =

2. The diagrams show three different distance–time graphs.

A

Distance / Time

B

Distance / Time

C

Distance / Time

Choose the **best** graph to fit each of these descriptions.

a Car moving at a constant speed: _____ [1 mark]

b Person running at a constant speed, then stopping and standing still:

_____ [1 mark]

c Person standing still: _____ [1 mark]

3. Look at the distance–time graph for a toy car.

a How far has the car moved after 4 seconds?

_____ [1 mark]

b How long does it take the car to move a total of 6 m?

_____ [1 mark]

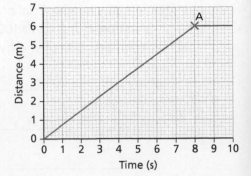

c Describe what happens at the point labelled 'A'.

The speed _____

The car _____ [2 marks]

4.

Practical

Carlos sets up an experiment to investigate a trolley rolling down a long slope. His measurements of time and distance are shown in the table.

Time (s)	Distance (m)
0	0
0.8	1.2
1.6	2.4
2.4	3.6
3.2	4.8

Look at the graph. The first three points have been placed on the graph.

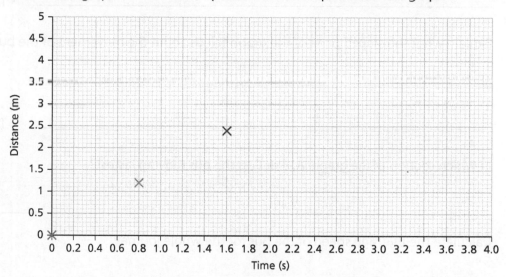

a Complete the graph by placing the two remaining points. [4 marks]

b Join all the points by drawing a line. [2 marks]

c Explain what the shape of the line tells us about the movement of the trolley.

_____

_____ [2 marks]

5

The graph shows the movement of a bus that travels between two different parts of town.

Challenge

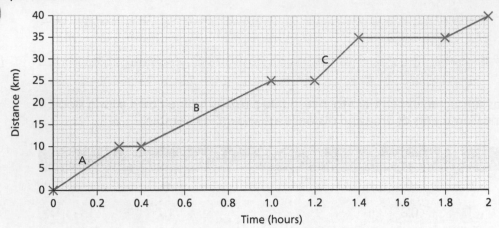

a Explain what the horizontal parts of the graph tell us about the movement of the bus.

_____

_____ [2 marks]

b Explain what the sloped parts of the graph tell us about the movement of the bus.

_____

_____ [2 marks]

c In which section of the graph, A, B or C, is the bus moving fastest?

_____ [1 mark]

d Calculate the **average** speed of the bus for the whole journey.

_____

_____

_____ [3 marks]

Self-assessment

Tick the column which best describes what you know and what you are able to do.

What you should know:	I don't understand this yet	I need more practice	I understand this
Rulers and measuring tapes can be used to measure distance			
Clocks, watches, stopwatches and light gates connected to electronic timers, data loggers or computers can be used to measure time			
Different pieces of apparatus give measurements with different levels of accuracy			
Average speed = distance travelled/time taken			
Speed is measured in metres per second (m/s)			
The gradient of a distance–time graph tells you the speed of a moving object			
The steeper the gradient of a distance–time graph, the faster an object is moving			

You should be able to:	I can't do this yet	I need more practice	I can do this by myself
Choose suitable apparatus to measure distance and time			
Identify important variables, choose which variables to change, control and measure			
Use a range of equipment accurately			
Present results as appropriate in tables and graphs			
Interpret data from secondary sources			
Make simple calculations			
Identify trends and patterns in results (correlations)			
Identify anomalous results and suggest improvements to investigations			
Discuss explanations for results using scientific knowledge and understanding			

If you have ticked 'I don't understand this yet' or 'I can't do this yet' or mostly 'I need more practice', have another look at the relevant pages in the Student's Book. Then make sure you have completed all the questions in this Workbook chapter and the review questions in the Student's Book. If you have already completed all the questions, ask your teacher for help and suggestions on how to progress.

Teacher's comments

Test-style questions

1. Lily investigates the speed of a ball rolling along a sloped track that is 1 m long.
She measures the distance using a measuring tape marked in centimetres.
She uses a stopwatch to time how long the ball takes to travel along the track.

The results are shown in the table.

Repeat number	Time taken (seconds)
1	0.83
2	0.78
3	0.91

a Which measuring device could Lily use to make her measurement of distance more accurate?

_____ [1 mark]

b Explain why using a stopwatch to measure the time is not very accurate, and suggest a better measuring device.

_____

_____

_____ [3 marks]

c Describe how the measurements Lily has made could be used to calculate the speed of the ball.

_____

_____ [2 marks]

2. A bus is travelling at a speed of 15 m/s. How long will the bus take to travel:

a 30 m?

_____ [1 mark]

b 600 m?

_____ [1 mark]

c 15 km?

_____ [1 mark]

3. Look at the two graphs showing the movement of two different objects.

Object 1

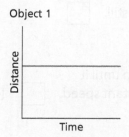

Object 2

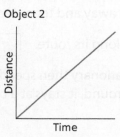

a Describe the **two** differences between the graphs.

_____

_____ [2 marks]

b Explain what the graphs tell us about the speed of each object.

_____

_____ [2 marks]

4. Look at the four graphs A to D.

A

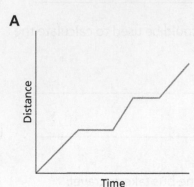

B

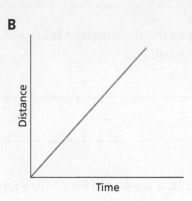

C

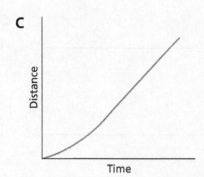

D

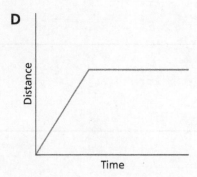

Match each graph to the description of the moving object. Choose the **best** description for each graph and write the letter of the graph next to the description.

(i) A car travelling at a steady speed. ☐

(ii) A person who walks a short distance away and then stands still. ☐

(iii) A bus that travels and makes stops along its route. ☐

(iv) An aircraft taking off, which starts stationary then speeds up until it rises into the air. After it leaves the ground, it stays at a constant speed. ☐ [4 marks]

5. Blessy sets up an experiment to measure how long it takes for a metal ball to roll down a track. The table shows the results.

Repeat number	Time taken (seconds)
1	2.9
2	3.2
3	4.0
4	2.9

a Explain why Blessy has run the experiment four times without changing the distance travelled.

_____

_____ [2 marks]

b Which result is anomalous (it does not fit a pattern)?

_____ [1 mark]

c Suggest **one** reason why the anomalous result could be wrong.

_____

_____ [2 marks]

d Choose the other **three** results. Calculate the average time taken.

_____

_____ [2 marks]

e Blessy measures the track and finds that it is 3.6 m long. Use your answer from (d) to calculate the average speed.

_____

_____ [2 marks]

10.1 Magnets and magnetic materials

Learning outcomes
- To describe the properties of magnets
- To investigate the properties of magnets and present results of investigations

1. Look at the diagrams. Choose which objects can be made into permanent magnets.

Tick (✓) the magnetic materials.

☐  iron bolt

☐  rubber eraser

☐  nickel block

☐  clear plastic rod

[2 marks]

2. Complete the sentences using the words from the list.

magnets poles south north east west permanent

Only some materials can be made into _____ magnets.

Magnetic _____ cannot exist on their own.

This means that every magnet must have a _____ pole and a

_____ pole. [4 marks]

3. Angelique writes some notes about magnets. She accidentally tore the page into pieces. She needs to stick the pieces back together.

repel each other

unlike magnetic poles

like magnetic poles

attract each other

Look at the pieces in the diagram and make **two** correct sentences about magnets. Write the sentences here.

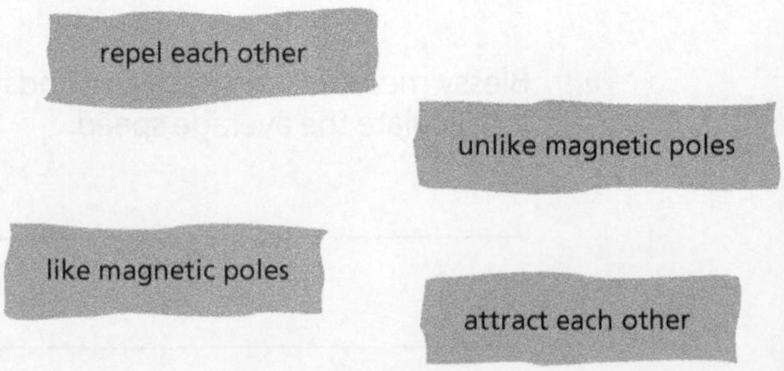

a _____

b _____ [2 marks]

4. Write a short explanation of what each term in the list means. The first one has been done for you.

Worked Example

a **Attract** means two objects that pull each other together. [1 mark]

Show Me

b **Repel** means two objects that _____

_____ . [1 mark]

> **Remember**
> Each explanation should be brief while saying exactly what each word means.

c **Magnetic pole** _____

_____ . [2 marks]

5. Look at the diagram of two magnets. Draw **two** arrows on the diagram to show the direction of the force acting on each magnet.

[2 marks]

6. Look at the diagram of two magnets. Draw **two** arrows on the diagram to show the direction of the force acting on each magnet.

[2 marks]

7. Look at the diagram of the magnetic field around a bar magnet. Complete the magnetic field lines.

[4 marks]

8. Look at the diagram of three bar magnets.

a Is the point marked 'A' a north pole or a south pole?

_____ [1 mark]

b Is the point marked 'B' a north pole or a south pole?

_____ [1 mark]

c Explain your answers to (a) and (b).

_____

_____

_____ [3 marks]

steel needle

9. Abhi has been given the apparatus shown in the diagram.

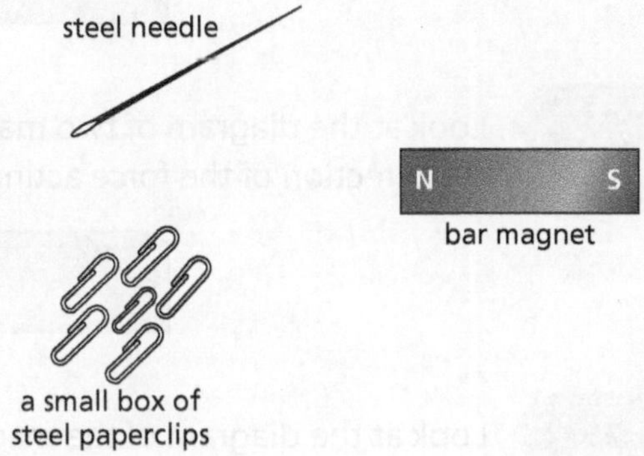

N   S

bar magnet

Challenge

Practical **a** Explain how Abhi should use the apparatus to show how the needle can be made to behave like a magnet.

a small box of steel paperclips

_____

_____

_____ [3 marks]

b Name the extra piece of apparatus Abhi would need to draw the magnetic field lines around a magnet.

_____ [1 mark]

c Explain why Abhi cannot use the needle as a permanent magnet.

_____

_____

_____ [2 marks]

10.2 Electromagnets

Learning outcomes
- To construct and use an electromagnet
- To identify important variables, choose which variables to change, control and measure

1. What is the relationship between a current in a wire and a magnetic field?

Tick (✓) the **best** answer from the list.

☐ A current flows only if a permanent magnet is placed nearby.

☐ A stationary magnet produces an electric current.

☐ A current is stopped from flowing if a permanent magnet is placed nearby.

☐ A current produces a magnetic field of its own. [1 mark]

2. Look at the diagram of an electromagnet. Complete the diagram by writing in the labels.

iron _____   wire_____

_____

_____

[4 marks]

3. Describe **two** important ways in which an electromagnet is different to a permanent magnet.

Show Me

A permanent magnet produces a magnetic field all the time.

An electromagnet produces a _____ _____ that

can be _____ .

The strength of the magnetic field around a permanent magnet

does not _____ .

We can _____ the strength of the magnetic field

around an _____ . [5 marks]

4. Choose materials suitable for making the core of an electromagnet.

Tick (✓) the **best** materials.

- [ ] Iron
- [ ] Polythene (plastic)
- [ ] Carbon
- [ ] Steel
- [ ] Mercury

[2 marks]

Remember
Always check your answers to make sure you have written the number of reasons a question asks you for. Do not write more reasons or fewer reasons.

The next **five** questions (5–10) are about an experiment to investigate the strength of an electromagnet. The diagram shows the apparatus used in the experiment.

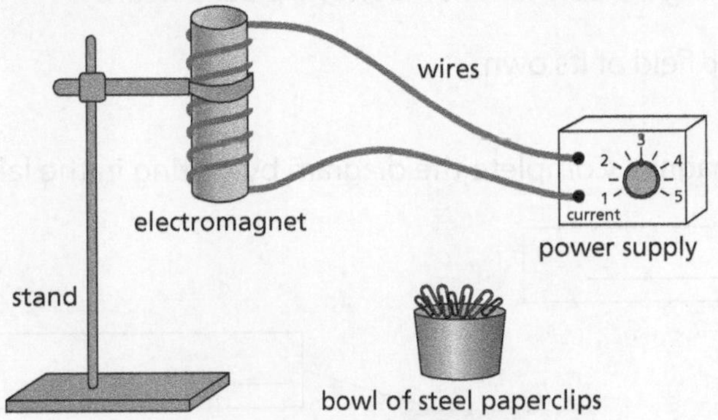

5. Practical

Jaina switches on the current and counts how many paperclips the electromagnet can pick up and hold. Jaina then repeats the experiment with different values of current.

Which type of variable is the **current** in this experiment? Tick (✓) the **best** answer.

- [ ] Independent variable
- [ ] Dependent variable
- [ ] Control variable
- [ ] Variable that does not affect the experiment

[1 mark]

6. Jaina produces a table to record the values of the variables. She adds a heading to each column that says what type of variable each value represents. Which variable is the **dependent** variable in this experiment? Tick (✓) the **best** answer.

- [ ] Current
- [ ] Number of paperclips
- [ ] Material the core is made from
- [ ] Number of turns in the coil

[1 mark]

7.

Practical

Jaina wants to know which variables she must control in the experiment.

Tick (✓) the variables that need to be controlled.

☐ Current

☐ Number of paperclips

☐ Material the core is made from

☐ Number of turns in the coil

[2 marks]

8.

Practical

Chen uses the same apparatus to investigate how changing the material in the core affects the strength of the electromagnet. The list shows the variables Chen knows about. For each variable, write whether it is the dependent variable, independent variable or a control variable in Chen's experiment. Write **dependent**, **independent** or **control**.

Current: _____

Number of paperclips: _____

Material the core is made from: _____

Number of turns in the coil: _____

[4 marks]

9.

Practical

Anastasia uses the same apparatus to measure the effect of another variable on the strength of the electromagnet. The table shows her results.

Write a short conclusion that describes Anastasia's results.

Current (A)	Number of paperclips lifted	Number of turns in coil	Material in core
2.0	2	6	iron
2.0	4	10	iron
2.0	6	14	iron
2.0	7	18	iron

_____

_____

_____ [3 marks]

10.

Challenge

Most motor vehicles include metals such as iron or steel in their bodies. When a vehicle stops working permanently or has come to the end of its useful life, it is scrapped. The vehicle is taken to a scrapyard where the different materials are separated out so they can be recycled.

The iron and steel are squashed into a cube shape to make them easier to move around. The cube can be picked up and moved by an electromagnet.

a Discuss the advantages of using an electromagnet to move these metals around. Describe **at least three** advantages and give reasons for your choices.

_____

_____

_____

_____

_____

_____ [6 marks]

b Describe **one** disadvantage of using an electromagnet.

_____

_____ [2 marks]

Self-assessment

∙∙

Tick the column which best describes what you know and what you are able to do.

What you should know:	I don't understand this yet	I need more practice	I understand this
Magnetic materials experience magnetic forces			
Every magnet has a north pole (N) and a south pole (S)			
Magnetic forces can attract or repel			
Like poles repel, unlike poles attract			
Magnetic field lines show the strength and direction of a magnetic field around a magnetised object			
When a current flows through a wire, it creates a magnetic field			
When a wire is coiled into a cylinder shape and a current is passed through it, it has a magnetic field like a bar magnet. This is called an electromagnet			

	I can't do this yet	I need more practice	I can do this by myself
The strength of an electromagnet is affected by the number of coils it has, the current and the type of core used to wrap the wires around			
Electromagnets have many uses because they can be turned on or off and their strength can be changed			

You should be able to:	I can't do this yet	I need more practice	I can do this by myself
Discuss the importance of developing empirical questions which can be investigated, collecting evidence, developing explanations and using creative thinking			
Select ideas and turn them into a form that can be tested			
Plan investigations to test ideas			
Identify important variables, choose which variables to change, control and measure			
Use a range of equipment correctly			
Present results in tables			
Present conclusions to others in appropriate ways			

If you have ticked 'I don't understand this yet' or 'I can't do this yet' or mostly 'I need more practice', have another look at the relevant pages in the Student's Book. Then make sure you have completed all the questions in this Workbook chapter and the review questions in the Student's Book. If you have already completed all the questions, ask your teacher for help and suggestions on how to progress.

Teacher's comments

Test-style questions

1. The diagrams show magnets in four different situations. Some of the magnetic poles are labelled, but some are not.

Complete the labels of the poles on each diagram.

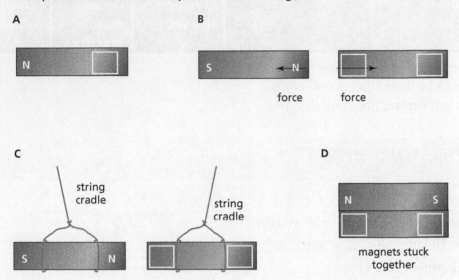

A

B

force   force

C

string cradle

string cradle

D

magnets stuck together

[4 marks]

2. The diagram shows two bar magnets placed close together. Join the dots using straight or curved lines to show the magnetic field around the two magnets. Two lines have been done for you.

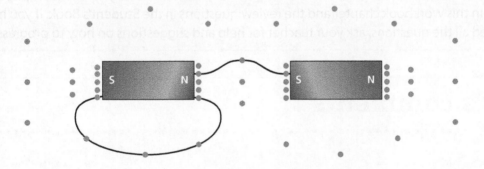

[6 marks]

3. Natalia investigates samples of different materials by moving a bar magnet slowly over each sample. Natalia lists her results in this table. Use her results to answer the questions.

Material	Plotting compass is placed near the sample [✓ = compass needle moves, ✗ = needle does not move]	Magnet is moved near the sample [✓ = sample pulled towards magnet, ✗ = no movement]	Plotting compass is placed near the sample, after the magnet is taken away [✓ = compass needle moves, ✗ = needle does not move]
Lodestone	✓	✓	✓
Glass	✗	✗	✗
Iron	✗	✓	✓
Coin made from steel and copper	✗	✓	✗

a Is lodestone a **magnetic** or **non-magnetic** material?

_____ [1 mark]

b Is glass a **magnetic** or **non-magnetic** material?

_____ [1 mark]

c Is iron a **magnetic** or **non-magnetic** material?

_____ [1 mark]

d Explain what happens to the iron sample at the different stages of the investigation.

_____

_____

_____ [3 marks]

e Is the coin a **permanent** magnet or a **temporary** magnet?

_____ [1 mark]

155

f Explain your answer to part (e).

_____

_____ [2 marks]

4. Drinks cans are usually made from metal. Some are made from steel. Others are made from aluminium. If we want to recycle the metals in drinks cans, we need to separate cans made from different metals.

Suggest a way of separating drinks cans. Explain the physics of your method.

_____

_____

_____ [3 marks]

5. Lucas has had an accident in his workshop. Two boxes of nails have split and become mixed up. Lucas has a bar magnet and thinks he can use it to lift the iron nails out of the mixture. Lucas finds the magnetic field is not strong enough to pick up one nail.

Lucas has an electrical toolbox that contains the components shown in the diagram. Explain to Lucas how he could use some components to make his bar magnet stronger.

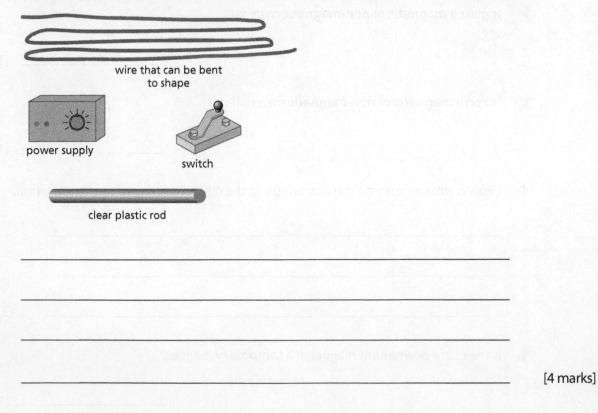

wire that can be bent
to shape

power supply

switch

clear plastic rod

_____

_____

_____

_____ [4 marks]